Nora Vanessa de Camp

New Methods: Brain electrophysiology in freely walking honeybees

Nora Vanessa de Camp

New Methods: Brain electrophysiology in freely walking honeybees

Brain recordings in social insects during real and virtual navigation: decision making, drugs, ripples and replay

Südwestdeutscher Verlag für Hochschulschriften

Impressum / Imprint
Bibliografische Information der Deutschen Nationalbibliothek: Die Deutsche Nationalbibliothek verzeichnet diese Publikation in der Deutschen Nationalbibliografie; detaillierte bibliografische Daten sind im Internet über http://dnb.d-nb.de abrufbar.
Alle in diesem Buch genannten Marken und Produktnamen unterliegen warenzeichen-, marken- oder patentrechtlichem Schutz bzw. sind Warenzeichen oder eingetragene Warenzeichen der jeweiligen Inhaber. Die Wiedergabe von Marken, Produktnamen, Gebrauchsnamen, Handelsnamen, Warenbezeichnungen u.s.w. in diesem Werk berechtigt auch ohne besondere Kennzeichnung nicht zu der Annahme, dass solche Namen im Sinne der Warenzeichen- und Markenschutzgesetzgebung als frei zu betrachten wären und daher von jedermann benutzt werden dürften.

Bibliographic information published by the Deutsche Nationalbibliothek: The Deutsche Nationalbibliothek lists this publication in the Deutsche Nationalbibliografie; detailed bibliographic data are available in the Internet at http://dnb.d-nb.de.
Any brand names and product names mentioned in this book are subject to trademark, brand or patent protection and are trademarks or registered trademarks of their respective holders. The use of brand names, product names, common names, trade names, product descriptions etc. even without a particular marking in this works is in no way to be construed to mean that such names may be regarded as unrestricted in respect of trademark and brand protection legislation and could thus be used by anyone.

Coverbild / Cover image: www.ingimage.com

Verlag / Publisher:
Südwestdeutscher Verlag für Hochschulschriften
ist ein Imprint der / is a trademark of
AV Akademikerverlag GmbH & Co. KG
Heinrich-Böcking-Str. 6-8, 66121 Saarbrücken, Deutschland / Germany
Email: info@svh-verlag.de

Herstellung: siehe letzte Seite /
Printed at: see last page
ISBN: 978-3-8381-3660-8

Zugl. / Approved by: Berlin, FU, Diss., 2012

Copyright © 2013 AV Akademikerverlag GmbH & Co. KG
Alle Rechte vorbehalten. / All rights reserved. Saarbrücken 2013

Summary		
Chapter	Title	Page
1. Summary		1
2. Introduction	hymenopteran insects	5
	optomotor behavior	6
	brain	7
	vision	9
	virtual environment	11
	operant learning	12
	ripples and replay	12
	octopamine, ritalin, cocaine	12
	references	22
3. Optomotor report	introduction	18
	material&methods	24
	results	25
	discussion	80
	references	82
4. New methods	introduction	84
	material&methods	87
	results	98
	discussion	119
	references	124
5. Are ripples/sharp waves older than the hippocampus?	introduction	129
	material&methods	133
	results	143
	discussion	176
	references	180
6. Drug effects on walking activity in the virtual environment.	introduction	187

	Summary	
	material&methods	190
	results	198
	discussion	225
	references	229
7.general discussion		236
	references	239

1.Abstract. The honeybee *Apis mellifera* is a eusocial insect. Only the queen and the drones are involved in sexual reproduction. The female worker bees show age related caste differentiation. Young bees are doing indoor works whereas older bees are flying outside the hive to collect pollen and nectar. Caste differentiation is regulated by diverse mechanisms, including biogenic amines and the distribution of receptors which are targets for neurohormones and neurotransmitters. Despite differences in the neurotransmitter system of insects (for example octopamine instead of noradrenalin) drugs like cocaine or methylphenidate, which are dopamine reuptake inhibitors in vertebrates, have similar behavioral and neural effects in bees as in vertebrates. Several comparative studies give rising evidence for the view, that basic concepts are conserved among protostomian and deuterostomian animals. The vertebrate pallium and the annelidan mushroom body develop from the same, molecularly defined subregion, octopaminergic reward neurons are resembling the properties of ventral tegmental dopaminergic neurons in the vertebrate brain and there are striking similarities in the molecular basis of learning and memory. Therefore it is not surprising that ripple-like potentials and replay are occuring in hymenopteran insects during virtual and real navigation. And most likely also in the last common ancestor of deutrostomes and protostomes, some 600 million years ago. Ripples might have a slightly differnt appearance in insects than in rodents but they are clearly associated with navigation and associated high frequency of place related firing neurons. This is an overwhelming evidence for the occurence of cognitive maps in insects, which have often been considered to be unnessesary to describe the insects navigational needs and therefore recognized as anti parsimonious.

1.Zusammenfassung. Die Honigbiene *Apis mellifera* ist ein eusoziales Insekt. Nur die Königin und die Drohnen sind an der sexuellen Fortpflanzung beteiligt. Weibliche Arbeitsbienen zeigen altersabhängige Kastendifferenzierung. Junge Bienen arbeiten im Stock, z.b. als Ammenbienen, wogegen ältere Bienen ausschwärmen um Nektar und Pollen zu sammeln. Die Kastendetermination unterliegt einer komplexen Regelung, u.a. durch biogene Amine und die Verteilung von Rezeptoren, die Ziel von Neurohormonen und Pheromonen sind. Trotz Unterschieden im Neurotransmitterhaushalt der Insekten (Oktopamin anstelle von Noradrenalin), haben Drogen wie Ritalin und Kokain, die Dopamintransporter in Vertebraten blockieren, ähnliche Effekte in Bienen, wie in Vertebraten.Vergleichende Studien geben vielseitige Hinweise auf konservierte, basale Mechanismen in Deuterostomiern und Protostomiern. Das Pallium der Vertebraten und der Pilzkörper der Anneliden entwickeln sich aus der gleichen, molekular definierten Untereinheit, octopaminerge Belohnungsneurone ersetzen dopaminerge Neurone im ventralen Tegmentum der Vertebraten und die molekularen Grundlagen von Lernen und Gedächtnis weisen deutliche Ähnlichkeiten auf. Aus dieser Perspektive ist es nicht verwunderlich, dass ripple-ähnliche Potentiale und replay während virtueller und realer Navigation in Hymenopteren auftreten. Höchstwahrscheinlich gab es diese Potentiale bereits im Gehirn der letzten gemeinsamen Vorfahren von Protostomiern und Deuterostomiern, vor 600 Millionen Jahren. Ripples mögen in Insekten einige abweichende Charakteristika als in Nagern aufweisen, sie sind jedoch klar mit hochfrequenten Phasen von Ortsneuronen gekoppelt und treten während virtueller und realer Navigation auf. Dieses Ergebnis ist ein deutlicher Hinweis auf kognitive Karten in Insekten, auch wenn dies in der Vergangenheit oft als nicht notwendige Annahme zur Erklärung der Navigationsleistungen von Insekten betrachtet wurde und somit als Verstoß gegen das Parsimonieprinzip gewertet wurde.

2.Introduction. The aim of this work is mainly the development of new methods to perform extracellular brain recordings in stationary walking insects. Since the construction of an immersive virtual environment is rather difficult, a second technique was established, which allows for extracellular recordings in freely walking insects of at least honeybee size. The second chapter gives a general introduction. The third chapter gives a short overview about pre experiments for optomotor behavior in walking bees. The fourth chapter focuses on the newly developed methods for extracellular recordings in walking bees, memory transfer of operant visual learning and neural correlates of decision making. The occurence of ripple like potentials, associated high unit activity in place related firing neurons and putative replay during sleep and navigation pauses in real and virtual navigation are the topics of the fifth chapter. The sixth chapter gives first insights to complex long-term changes in network spiking activity during cocaine and methylphenidate treatment in running bees and bumblebees. The new findings are discussed in the seventh chapter. The last part is the released utility patent at dpma.

Hymenopteran insects: Honeybees, bumblebees and hornets.

Honeybees (*Apis mellifera*) are eusocial insects with a reproductive division of labor. The queen is laying eggs and is fertilized by the drones. The female worker bees are not active in reproduction. During their ontogenesis they are working in different castes inside the hive or outside. Young female worker bees are starting with in-hive jobs like nursing. Older bees are flying outside the hive to collect pollen and nectar. The system of caste differentiation is very complex and regulated by diverse factors like pheromones, neurotransmitters and neurohormones. Bumblebees show the full scale of social development, depending on the species. Some species are eusocial with a reproductive division of labor. Bumblebees build smaller colonies than honeybees. Hornets are predatory social insects. All mentioned species are haplodiploid, which means, that females develop from fertilized eggs and are therefore generally diploid, whereas males develop from unfertilized eggs and are haploid. A common effect is, that daughters which are related through both parents show an unusually high degree of relatedness (¾). Hamilton (1964) was the first who used this phenomenon as an ultimate explanation for the development of eusociality and reproductive division of labor. Notably, diploid hornet drones have been observed (Foster et al., 2000).

Optomotor behavior. The typical optomotor response is characterized by an alignment of the body axes during flight in the same direction as a surrounding, moving pattern of gratings (Reichardt, 1969). This behavior has been confirmed for several species, including bees (Srinivasan, 1991). Interestingly, this behavior is less clear in walking animals. It is assumed, that the visual stumulus alone is not sufficient, but additional, especially wind stimuli are needed to elicit an optomotor response (Haag et al., 2010). Blowflys show optomotor head movements only in special gain states, which are characterized by head jittering (Rosner et al., 2009)

Brain. The insect brain consists of three parts. The protocerebrum includes the mushroom body, protocerebral lobes including the optic lobes and the central complex (Goll, 1967). The deutocerebrum consists mainly of the antennal nerve and lobes (Goll, 1967). The third part, the tritocerebrum, developed later than the former two parts. Mayer et al. (2010) found that the Onychophora as sister group of all other Artropoda have no tritocerebrum. Taste perception neurons are projecting to the tritocerebrum (Thorne et al., 2004). The mushroom body is a protocerebral brain structure which has been shown to be involved in learning and memory (Menzel et al., 1974, Erber et al., 1980). Every mushroom body consists of approximately 170 000 intrinsic Kenyon cells (Kenyon, 1896) (Witthöft, 1967). The Kenyon cell dendrites give rise to the medial and lateral calyx of the mushroom body, which have been identified as the main input region for the mushroom body intrinsic neurons (Schürmann, 1974). The Kenyon cell axons are running in two colums via the pedunculus and finally split in two parts, the alpha lobe (vertical lobe) and beta lobe (medial lobe) (Kenyon, 1896). The three calycal neuropils receive different sensory modalities. The lip region receives olfactory information from the projection neurons, the collar region receives input from visual neurons and the basal ring receives input of intermixed sensory modalities (Mobbs, 1982). This modality specific topography is conserved throughout the mushroom body and visible as a layered pattern in the alpha and beta lobes

(Mobbs, 1982, Rybak, 1994). Mushroom body extrinsic neurons have their dendrites in the pedunculus, alpha or beta lobes and connect the mushroom body to other neuropile regions (Mobbs, 1982, Schürmann, 1987, Rybak&Menzel, 1993). Mushroom body extrinsic neurons are often sensitive for a variety of different sensory stimuli (Erber, 1978, Homberg&Erber, 1979, Gronenberg, 1987, Mauelshagen, 1993). This decomposition of sensory modalities via mushroom body extrinsic neurons has been interpreted as sensory integration (Erber, 1987).

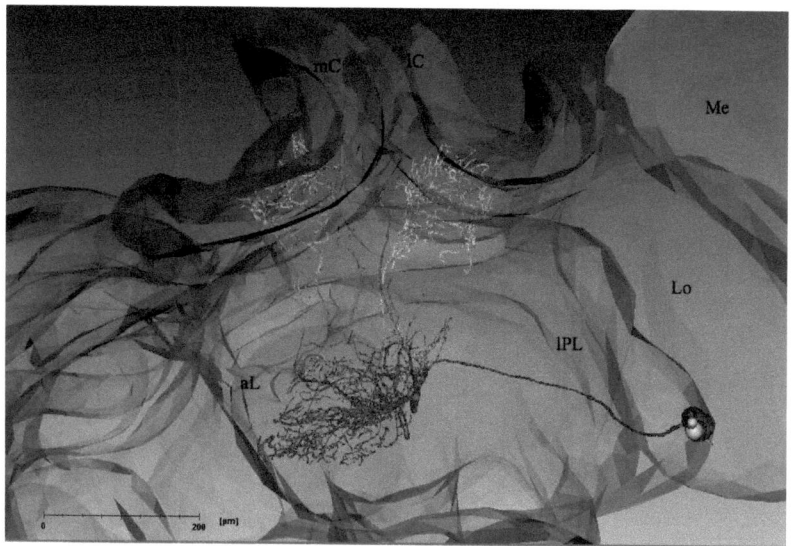

Fig.I1: A reconstructed feedback neuron of the protocerebro calycal tract (PCT) in the standard brain atlas of the honeybee (de Camp, 2009). The soma (beige) is located in the lateral protocerebral lobe (lPL) at the border to the optic lobes (Lo=Lobula, Me=Medulla).The primary neurite (red) is spanning the protocerebral lobe towards the alpha lobe (aL) of the mushroom body. This PCT neuron has dendritic arborizations in the alpha lobe (magenta) as well as in the deeper beta lobe (dendrites in orange). An axon (green) connects the dendritic regions in the alpha and beta lobe with dendritic arborizations in the lateral Calyx (lC, dendrites in yellow) and the medial calyx (mC, dendrites in cyan).

Vision. Since honeybees are collecting pollen and nectar from flowers it is not surprising that they are able to discriminate colours, as it has been shown by Karl von Frisch (1914). The spectral discrimination has three peaks: 345nm (UV), 440nm (blue) und 550nm (green) (Thomas&Autrum, 1965). The trichromtatic vision has been confirmed in learning experiments with honeybees (von Helverson, 1972) and for hornets and bumblebees via intracellular recordings (Peitsch et al., 1992). Hornets are flying in dim light conditions without having anatomical adaptations of the apposition eye (Kelber et al., 2011). The compound eye consists of several ommatidia. Each ommatidium has a cornea and a crystal cone, which together build up a functional lens that projects the light onto the rhabdomere. The rhabdomeres are the sites of phototransduction. They carry the light sensitive photopigments (rhodopsin which is converted to metarhodopsin by light absorption) and transduce the light perception into a neural signal (Stavenga, 2002). Bees and other hymenoptera have an apposition eye. The rhabdomeres of the nine photoreceptors are fused in one optical waveguide (Stavenga, 2002). In contrast to former results, which reduce motion vision to the blue-green receptors in insects (Joesch et al., 2010) in *Drosophila* the color photoreceptors R7 and R8 converge with and change R1-R6 (blue-green receptors) as well as associated large monopolar cell ouputs and thereby improve motion discrimination (Wardill et al., 2012). Optic comissures have been described which detect movements and connect both lobulae, as well as visual interneurons which are projecting to higher order brain centers (Hertel et al., 1987). The cognitive capacities of honeybee visual processing are a recent issue of debate (Horridge, 2004, Dyer, 2012). A retinotopic template-matching mechanism for pattern discrimination, as it has been suggested for *Drosophila*, could not have been confirmed for honeybees (Efler&Ronacher, 2000). Dyer (2012) proposes multiple mechanisms of pattern recognition.

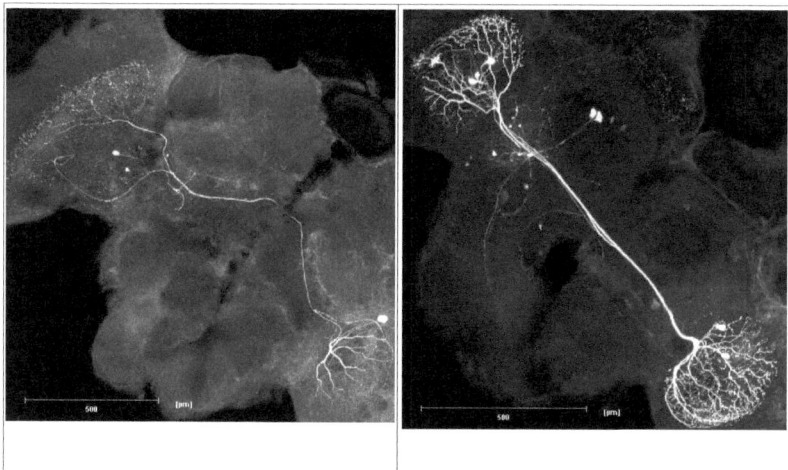

Fig.I2: Optical interneurons in the bee brain (de Camp, 2009). Two examples for neurons which are connecting the optic lobes of both brain hemispheres. The scale indicates 500μm. The ocelli (dorsal) are located in the upper right edge, respectively. The subesophageal ganglion (SOG, ventral part of the brain) is located in the bottom left edge, respectively.

Virtual environment. Virtual environments are used to study navigation in humans and animals. Usually virtual navigation is stationary, allowing for further measurments in the subjects, for example intracellular recordings (Harvey et al., 2009) or imaging (Dombeck et al., 2010). A second advantage of visual virtual environments is the extended landscape which can be presented in comparison with the lab situation. It has been shown, that the visual feedback is one important parameter to create immersive properties (Keller et al., 2012). Nevertheless, mice are able to learn and navigate in the virtual reality (Harvey et al., 2009).

Operant learning. In honeybees, classical olfactory conditioning, using the proboscis extension response (PER) is a widely used and a reproducable method to study learning in an insect model system (Kuwabara, 1957). Nevertheless, classical conditioning of colour cues is hardly possible with the PER paradigm (Niggebrügge et al., 2009). New results suggest, that visual classical conditioning is age dependent (Dobrin et al., 2012). So far, operant conditioning was used in restrained bees with an antennal paradigm (Haupt, 2007). Since operant learning requires initial actions of the animals, which can be associated with an rewarding or punishing outcome, operant conditioning in stationary walking honeybees is a first approach to combine this form of learning with electrophysiology. For *Drosophila*, operant conditioning is well established. The animals are flying stationary with a torque meter which allows for a control of the surrounding visual cues in the flight arena (VanSwinderen, 2003).

Ripples and Replay. In rodents it has been shown, that sharp waves in the local field potential are occuring during pauses in navigation or slow wave sleep (Buzsáki et al., 1992). Furthermore, ripples are associated with high unit activity of hippocampal place cells, which have been activated during active navigation (Davidson et al., 2009). A disruption of ripple events and the associated replay of place cell encoded spatial representations leads to impaired place memory consolidation (Girardeau et al., 2009).

Octopamine, Ritalin, Cocaine. Cocaine, a alcaloid of the coca plant and methylphenidate (Ritalin), an amphetamine derivate are involved in dopamine reuptake inhibition in vertebrates and insects (Ritz et al., 1987,

Corey et al., 1994). In insects, cocaine blocks dopamine as well as octopamine transporters (Gallant et al., 2003). It has been shown that cocaine has influences on the honeybee dance behavior (Barron et al., 2009). Methylphenidate rescued some attention-like deficits in Drosophila memory mutants (VanSwinderen&Brembs, 2009). Ritalin is a widely used drug to treat hyperactivity disorder in children (Accardo&Blondis, 2001, Frölich et al., 2012). Recently it has been shown, that Ritalin has severe influences on the neural activity in the nucleus accumbens in chronically treated rats (Chong et al., 2012). Octopamine is a neurotransmitter which is predominantly occuring in invertebrates. It resembles some structural and functional features with noradrenalin. For example, octopamine is involved in stress responses in insects (Roeder, 1999).

References.

Accardo, P., Blondis, T.A. (2001) What's all the fuss about Ritalin? J. Pediatr. **138:** 6–9.

Barron, A.B., Maleszka, R., Helliwell, P.G., Robinson, G.E. (2009) Effects of cocaine on honey bee dance behaviour. J. Exp. Biol. **212:** 163-168.

Buzsáki, G., Horváth, Z., Urioste, R., Hetke, J., and Wise, K. (1992). High frequency network oscillation in the hippocampus. Science **256:** 1025–1027.

de Camp, N. (2009) Struktur und Funktion der PCT Neurone im Bienengehirn. Diploma thesis, FU Berlin.

Chong, S.L., Claussen, C.M., Dafny, N. (2012) Nucleus accumbens neuronal activity in freely behaving rats is modulated following acute and chronic methylphenidate administration. Brain Research Bulletin **87:** 445–456.

Corey, J. L., Quick, M. W., Davidson, N., Lester, H. A. and Guastella, J. (1994). A cocaine-sensitive *Drosophila* serotonin transporter – cloning, expression, and electrophysiological characterization. Proc. Natl. Acad. Sci. USA **91:** 1188-1192.

Davidson, T.J., Kloosterman, F., Wilson, M.A. (2009) Hippocampal Replay of Extended Experience. Neuron **63:** 497-507.

Dobrin, S.E., Fahrbach, S.E. (2012) Visual associative learning in restrained honey bees with intact antennae. PloS One 7(6):e37666. Epub 2012 Jun 6.

Dombeck, D.A., Harvey, C.D., Tian, L., Looger, L.L., Tank, D.W. (2010) Functional imaging of hippocampal place cells at cellular resolution during virtual navigation. Nat. Neurosci. **13:** 1433-1440.

Dyer, A.G. (2012) The mysterious cognitive abilities of bees: why models of visual processing need to consider experience and individual differences in animal performance. J Exp Biol **215:** 387-395.

Efler, D., Ronacher, B. (2000) Evidence against a retinotopic-template matching in honeybees'
pattern recognition. Vision Research **40**: 3391–3403.

Erber, J. (1978) Response characteristics and after effects of multimodal neurons in the mushroom body area of the honey bee. Physiol.Entomol. **3**: 77-89.

Erber, J., Masuhr, T., Menzel, R. (1980) Localization of short-term memory in the brain of the bee, *Apis mellifera*. Phys Entomol **5**: 343-358.

Erber, J., Homberg, U., Gronenberg, W. (1987) Functional roles of the mushroom bodies in insects. In: Gupta AP (ed) Arthropod brain: Its evolution, development, structure and functions. John Wiley and Sons, New York, pp485-511.

Foster, K.R., Ratnieks, F.L.W., Raybould, A.F. (2000) Do hornets have zombie workers? Molecular Ecology **9**: 735–742.

Frisch, K. von (1914) Der Farbensinn und Formensinn der Biene. Zool. Jahrb. Abt. Allgem. Zool. Physiol. **35**: 1-188.

Frölich, J., Banaschewski, T., Spanagel, R., Döpfner, M., Lehmkuhl, G. (2012) Die medikamentöse Behandlung der Aufmerksamkeitsdefizit-Hyperaktivitätsstörung im Kindes- und Jugendalter mit Amphetaminpräparaten. Zeitschrift für Kinder- und Jugendpsychiatrie und

Psychotherapie, **40 (5):** 287–300.

Gallant, P., Malutan, T., McLean, H., Verellen, L., Caveney, S. and Donly, C. (2003) Functionally distinct dopamine and octopamine transporters in the CNS of the cabbage looper moth. Eur. J. Biochem. **270:** 664-674.

Girardeau, G., Benchenane, K., Wiener, S.I., Buzsáki, G., Zugaro, M. B. (2009) Selective suppression of hippocampal ripples impairs spatial memory. Nat. Neurosci., **12:** 1222-1223.

Goll, W. (1967) Strukturuntersuchungen am Gehirn von *Formica*. Z.Morph. Ökol. Tiere **59:** 143-210.

Gronenberg, W. (1987) Anatomical and physiological properties of feedback neurons of the mushroom bodies in the bee brain. Exp. Biol **46:** 115-125.

Hamilton, W. D. (1964) The Genetical Evolution of Social Behaviour. I. J. Theoret. Biol. **7:** 1-16.

Haupt, S.S. (2007) Central gustatory projections and side-specificity of operant
antennal muscle conditioning in the honeybee. J Comp Physiol A **193:** 523–535.

Harvey, C.D., Collman, F., Dombeck, D.A., Tank, D.W. (2009) Intracellular dynamics of hippocampal place cells during virtual navigation. Nature **461**: 941-946.

Hertel, H., Schäfer, S., Maronde, U. (1987) The physiology and morphology of visual commisures in the honeybee brain. J.exp.Biol. **133**: 283-300.

Hertel, H., Maronde, U. (1987) The Physiology and Morphology of Centrally Projecting Visual Interneurones in the Honeybee Brain. J Exp Biol **133**: 301-315.

Homberg, U. and J. Erber (1979) Response Characteristics and Identification of Extrinsic Mushroom Body Neurons of the Bee. Z.Naturforsch. **34**: 612-615.

Kelber, A., Jonsson, F., Wallén, R., Warrant, E., Kornfeldt, T., Baird, E. (2011) Hornets Can Fly at Night without Obvious Adaptations of Eyes and Ocelli. PloS ONE 6(7): e21892. doi:10.1371/journal.pone.0021892.

Keller, G.B., Bonhoeffer, T., Hübener, M. (2012) Sensorimotor mismatch signals in primary visual cortex of the behaving mouse. Neuron **74(5)**: 809-15.

Kenyon, F.C. (1896) The meaning and structure of the so-called "mushroom bodies" of the hexapod brain. Am. Nat. **30**: 643–650.

Kuwabara, M. (1957). Bildung des bedingten Reflexes von Pavlovs Typus bei der
Honigbiene, Apis mellifica. J. Fac. Sci. Hokkaido Univ. Ser. VI Zool. **13:** 458-464.

Mayer, G., Whitington, P.M., Sunnucks, P., Pflüger H.-J. (2010) A revision of brain composition in *Onychophora* (velvet worms) suggests that the tritocerebrum evolved in arthropods. BMC Evolutionary Biology **10:** 255.

Mauelshagen, J. (1993) Neural correlates of olfactory learning paradigms in an identified neuron in the honeybee brain. J. Neurophysiol. **69(2):**609-625.

Menzel, R., Erber, J., Masuhr, T. (1974) Learning and memory in the honeybee. In: Barton Browne L (ed) Experimental analysis of insect behavior. Springer, Berlin, Heidelberg, New York, pp 195-217.

Mobbs, P.G. (1982) The brain of the honeybee *Apis mellifera*. I. The connections and spatial organization of mushroom bodies. Phil. Trans. R. Soc. Lond. B **298:** 309-354.

Niggebrügge, C., Leboulle, G., Menzel, R., Komischke,B., Hempel de Ibarra, N. (2009) Fast learning but coarse discrimination of colours in restrained honeybees. J. Exp. Biol. **212:** 1344-1350.

Peitsch, D., Fietz, A., Hertel, H., de Souza, J., Fix Ventura, D., Menzel, R.

(1992) The spectral input systems of hymenopteran insects and their receptor based colour vision. J Comp Physiol A **170**: 23-40.

Reichardt, W. (1969). Movement perception in insects. In Processing of optical data by organisms and by machines. (Edited by Reichardt W.). Academic Press, New York.

Ritz, M.C., Lamb, R.J., Goldberg, S.R., Kuhar, M.J. (1987) Cocaine receptors on dopamine transporters are related to self-administration of cocaine. Science **237(4819):** 1219-23.

Roeder, T. (1999) Octopamine in invertebrates. Prog. Neurobiol. **59:** 533–561.

Rosner, R., Egelhaaf, M., Grewe, J., Warzecha, A.K. (2009) Variability of blowfly head optomotor responses. J. Exp. Biol. **212**, 1170-1184.

Rybak, J. (1994) Die strukturelle Organisation des Pilzkörpers und die synaptische Konnektivität protocerebraler Interneurone im Gehirn der Honigbiene, *Apis mellifera*. Dissertation, FU Berlin.

Rybak, J., Menzel, R. (1993) Anatomy of the mushroom bodies in the honey bee brain: the neuronal connections of the alpha lobe. J. Comp. Neurol. **334:** 444-465.

Schürmann, F.-W. (1987) The architecture of the mushroom bodies and related neuropiles in the insect brain. In: Gupta AP (ed) Arthropod brain: Its evolution, development, structure and functions. John Wiley and Sons, New York, pp231-256.

Schürmann, F.-W. (1974) Bemerkungen zur Funktion der Corpora pedunkulata im Gehirn der Insekten aus morphologischer Sicht. Exp. Brain. Res. **19:** 406-432.

Srinivasan, M.V., Lehrer, M., Kirchner, W.H., Zhang, S.W. (1991) Range perception through apparent image speed in freely flying honeybees. *Visual Neuroscience* **6:** 519-535.

Stavenga, D.G. (2002) Colour in the eyes of insects. J Comp Physiol A **188:** 337–348.

Thorne, N., Chromey, C., Bray, S., Amrein, H. (2004) Taste Perception and Coding in *Drosophila*. Current Biology **14:** 1065–1079.

Witthöft, W. (1967) Absolute Anzahl und Verteilung der Zellen im Hirn der Honigbiene. Z. Morphol. Tiere **61:** 160-184.

Van Swinderen, B., Brembs, B. (2009) Attention-Like Deficit and Hyperactivity in a *Drosophila*
Memory Mutant. J. Neurosci. **30(3):**1003–1014.

Van Swinderen, B., Greenspan, R.J. (2003) Salience modulates 20-30 Hz brain activity in *Drosophila*. Nat. Neurosci. **6:** 579-586.

3. Optomotor report: Bee vs Ant

Abstract. The optomotor response in tethered walking insects appears to be much weaker and less uniform than in tethered flying insects. The aim of this study was to examine, if the absence of optomotor walking behavior in a virtual environment might be due to the projected image quality or other setup specific errors. Therefore, the experiments presented here were performed with a drum, which had a stripe pattern on its inner surface. It was recently shown that some proprioceptive inputs during flight are enhancing the optomotor response as well as inner gain states. In the following, the optomotor behavior of tethered walking ants and bees is compared.

Introduction. The disadvantage of poor stereopsis-based depth perception of the insect compound eye, which is a result of the short distance between the eyes and their comparatively low spatial resolution is overcome by self induced, translational, lateral image flow (Lehrer, 1996, Srinivasan&Zhang, 2004). Therefore depth perception in insects depends on behaviorally acquired motion parallax information, like the "turn-back-and-look behavior in honeybees (Lehrer et al., 1991) or zig-zag flights in wasps, which produce high differential image motion and changes in the direction of motion vectors especially at edges or in near vicinity to them (Voss&Zeil, 1998). Since honeybees, flying in a continuously rotating, striped drum do not turn upside down, but hold the head at a certain angle, it is clear that the visually guided gaze stabilization is limited by other factors like the dorsal light response or gravity perception (Boeddeker&Hemmi, 2010).

As shown by Reichardt (1969) insects which are flying tethered in a striped drum tend to follow the stripe rotation. If the drum rotates clockwise the insect will generate a yaw torque in the same direction if the drum is rotating in counterclockwise direction the animal will generate yaw torques in the counterclockwise direction. A flicker frequency of 8 Hz is most effective in eliciting optomotor responses in tethered flying honeybees (Kunze, 1961). Furthermore, illumination was positively correlated with the optomotor response. A correlation between illumination and flicker frequency exists in bees: Higher levels of illumination shift the optimal flicker frequency (with respect to the strength of optomotor response) to higher values (Kunze, 1961).

But what about tethered walking animals? The results of studies in this field are less uniform.

Blowflies show optomotor head movements only in special gain states, characterized by head jittering (Rosner et al., 2009). Head jitter movements go along with high optomotor gain state. Nevertheless animals in the same state showed huge variability in optomotor responses which was independent of haltere reafferents and therefore possibly central influenced (Rosner et al., 2009). The optomotor response in *Drosophila* is modulated by olfactory inputs in a mushroom body-dependent way (Chow et al., 2011).

Haag et al. (2010) could show that optomotor responses in blowflies depend on inputs to the haltere associated campaniform sensillae and the Johnston organ on the antennae. The visual information from a central neuron alone is not sufficient to elicit an optomotor response by depolarizing the ventral cervical nerve motoneuron (VCNM). Only additional activity of neurons from the wind sensitive Johnston organ and the campaniform sensillae led to a depolarization of the VCNM above threshold and the resulting neck movement. The activity of neck motor neurons and associated visual responses in the honeybee are modulated by inputs from the dorsal ocelli (Hung et al., 2011).

Srinivasan et al. (1991) observed optomotor behavior in naive flying bees who learned to fly through a tunnel with a stripe pattern on one side but as training proceeds the optomotor behavior was suppressed. These findings are in contrast to similar experiments with flys.

Material and Methods. The Bees (n c.a. 5, *Apis mellifera*) were caught from the winter indoor hive and anesthetized by cooling on ice for the fixation of an insect needle on top of the thorax with dental wax.

Ants (n c.a. 5, *Pachycondyla, Myrmecia)* were clamped with insect needles to prevent moving. The thorax fixation was the same as for bees.

The Setup was a Styrofoam ball, floating on an air stream and a surrounding barrel with black stripes on white ground at the inner surface. The black stripes had a width of 4 cm as well as the white spaces in between. The inner surface of the barrel had 15 black and 15 white stripes and a diameter of approximately 41 cm. The circumference of the barrel was 132 cm at the place where the animal was positioned and 115 cm at a position 60 cm higher at the bottom (during the experiment the top of the barrel - it was hanging upside down). At the closed side of the barrel a pivotable hanger assembly was mounted. This allowed manual rotation of the barrel around the fixed animal. The resulting frequency of stripe rotation was approximately 5Hz.

Analysis. Matlab 2007a (simulink) was used as well as excel and iworks. For statistical analysis the total time of rotation movements was detected with matlab and normalized as ratio between the single direction rotation time and the full amount of rotation time for one animal and one experiment.

Every experiment had a duration of three minutes. The pause between two experiments took at least 3 minutes for the same animal. The three different versions of optomotor tasks were 3 minutes continous clockwise rotation, 3 minutes continous counterclockwise rotation and one minute right, one minute left and again one minute right stripe rotation.

For single experiment analysis the percentage of rotation events was measured as difference between two consecutive mouse updates. If the difference of two consecutively measured rotation values was negative, the animal rotated left. If it was positive, the animal rotated right. The mouse update time varied from 20 to 400 ms but in random order so this measuring artifact was the same for any direction of rotation movement. To eliminate possible tiny rotation artifacts by the floating ball itself, the percentage was calculated with a threshold of $0.01°$ for rotation values as well.

The control experiments took place in the same environment as the optomotor tasks but the barrel was not rotated. The rotation movements by the animals were measured for three minutes with stationary stripes. The walking data were collected with a program written by Sören Hantke. The ants were provided by Dirk Drenske and data collection took place together with Dirk Drenske.

Results. As shown in figure 1 bees and ants are running more or less in random direction. The standardized mean Norm Pro is the amount of time for running in the same direction as the gratings divided by the total amount of time in which rotation took place.

The difference from 0,5 (50% running time in the one direction and 50% running time in the other direction) is never as big as the standard deviation, neither for bees nor for ants.

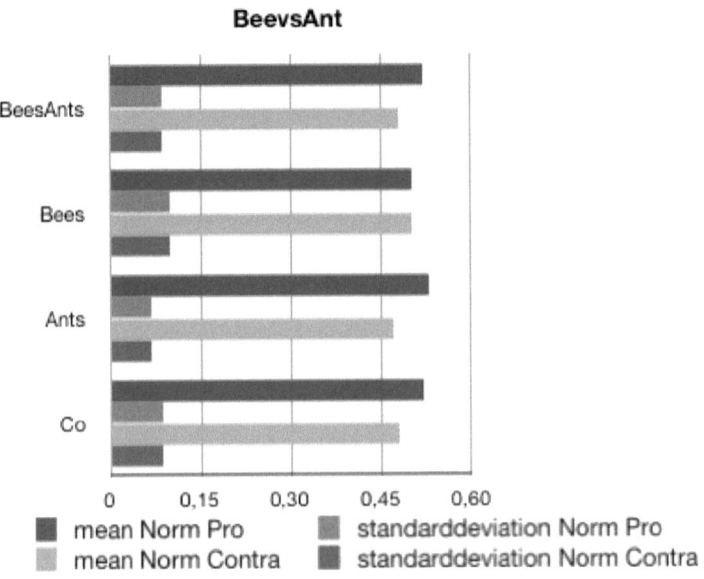

Fig.1: Overview for all experiments. The Blue bars represent the mean of the standardized (see methods) time for Pro (running in the same direction as the stripe rotation). The yellow bars represent the standardized time for Contra (running in the opposite direction as the stripe rotation). Green and red are showing the standard deviation, respectively. The standardized values must be in the range from zero to one. With zero meaning no time and one meaning maximal time.

In the first row, bees and ants of clockwise, counterclockwise and right-left-right experiments are pooled. Beneath Bees and Ants are shown as separate groups for the same experiments and at the bottom the controls are plotted for comparison.

Results for right-left-right-experiments:

The experiments last for 3 min. For the first and last minute, stripes rotate clockwise. In between the stripes rotate for one minute in counterclockwise direction. For the first and the last minute Pro means clockwise rotation of the animal. For the second minute Pro means counterclockwise rotation of the animal.

In the following histograms, rotation is shown as percentage of rotation events. Every rotation event is measured as difference between two following rotation values, updated by the computer mice. This update time is irregular and in the range of 20 to 400 ms. If the difference is negative the rotation is left oriented, if it is positive the rotation is oriented to the right. It is only shown the direction in context to the stripe rotation. If animal and stripe rotation are in the same direction it is called "Pro". If animal and stripe rotation are opposing it is called "Contra". There are no significant results for one preferred direction, some bees are running more "Contra" others more "Pro".

The second picture is always showing the rotation history during the whole experiment. Stripe rotation is color coded with magenta meaning left or counterclockwise rotation and cyan meaning right or clockwise rotation. The animal rotation is plotted in degree per time.

Bees:

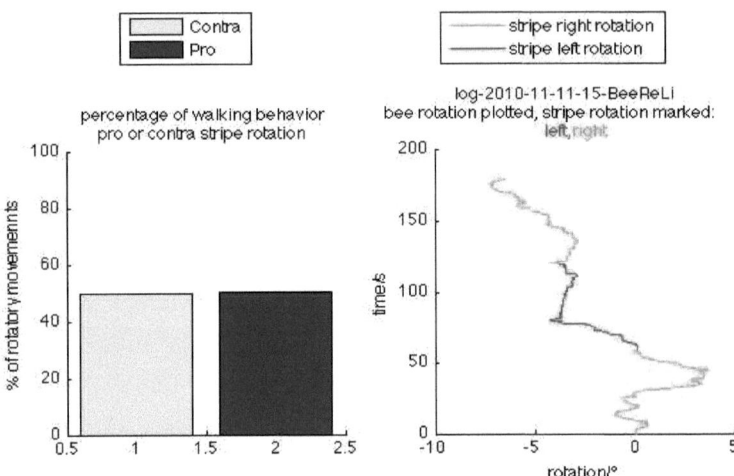

Fig.2: On the left, the percentage of walking behavior pro or contra stripe rotation is plotted. This means, not total time but single right or left rotation events, measured as difference between two consecutive mouse update events (see methods). Units on the x-axis are not meaningful. On the right side, the rotatory run is shown. For the first minute - right rotation of the stripes - it is shown in cyan. The second minute, with left rotation of the stripes, is color coded in magenta. Cyan is again marking right rotation of the stripes during the third minute. There is no significant correlation between walking behavior and stripe rotation.

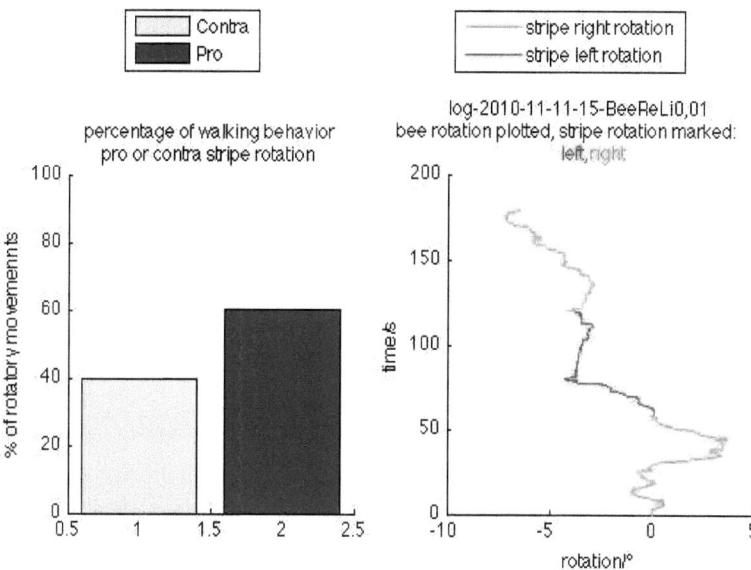

Fig.3: On the left, the percentage of walking behavior pro or contra stripe rotation is plotted. This means, not total time but single right or left rotation events, measured as difference between two consecutive mouse update events (see methods). Units on the x-axis are not meaningful. On the right side, the rotatory run is shown. For the first minute -a right rotation of the stripes- it is shown in cyan. The second minute, with left rotation of the stripes is color coded in magenta. Cyan is again marking right rotation of the stripes during the third minute. The figure is the same as no 2 but with a threshold of 0.01 for rotatory data (see methods) in the histogram.

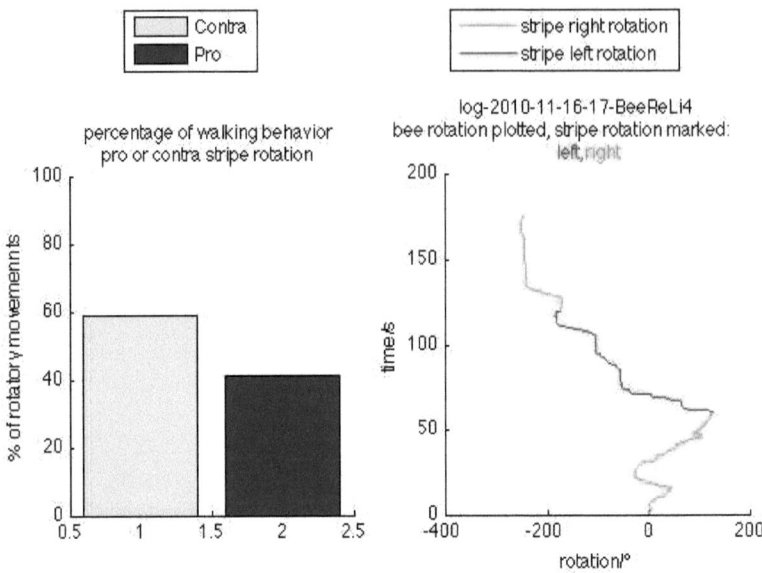

Fig.4: On the left, the percentage of walking behavior pro or contra stripe rotation is plotted. This means, not total time but single right or left rotation events, measured as difference between two consecutive mouse update events (see methods). Units on the x-axis are not meaningful. On the right side, the rotatory run is shown. For the first minute -a right rotation of the stripes- it is shown in cyan. The second minute, with left rotation of the stripes is color coded in magenta. Cyan is again marking right rotation of the stripes during the third minute. There is no significant correlation between walking behavior and stripe rotation.

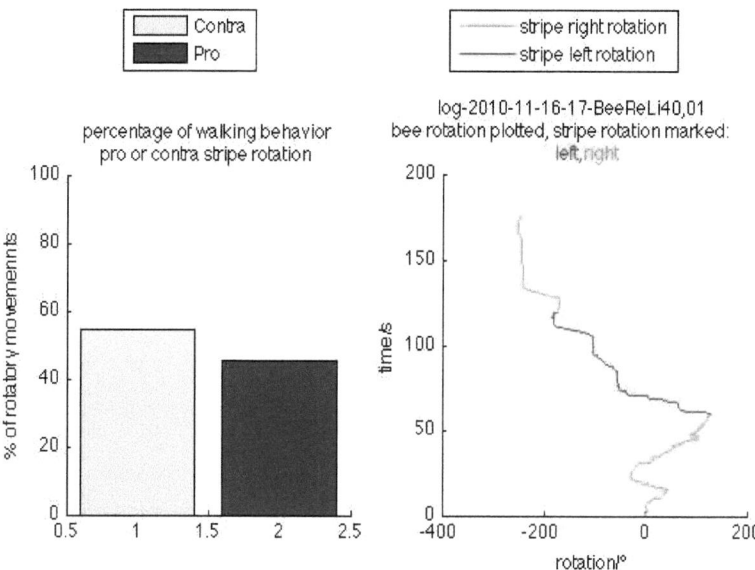

Fig.5: On the left, the percentage of walking behavior pro or contra stripe rotation is plotted. This means, not total time but single right or left rotation events, measured as difference between two consecutive mouse update events (see methods). Units on the x-axis are not meaningful. On the right side, the rotatory run is shown. For the first minute - right rotation of the stripes - it is shown in cyan. The second minute, with left rotation of the stripes is color coded in magenta. Cyan is again marking right rotation of the stripes during the third minute. The figure is the same as no 2 but with a threshold of 0.01 for rotatory data (see methods) in the histogram.

Ants:

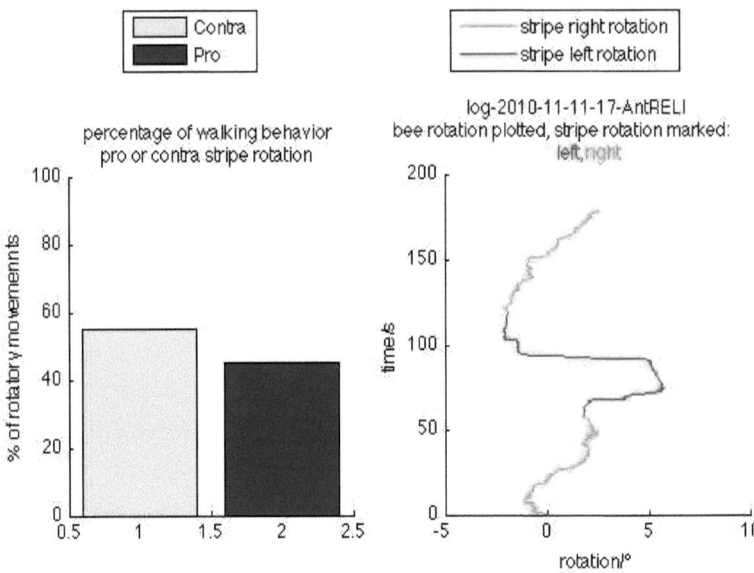

Fig.6: On the left, the percentage of walking behavior pro or contra stripe rotation is plotted. This means, not total time but single right or left rotation events, measured as difference between two consecutive mouse update events (see methods). Units on the x-axis are not meaningful. On the right side, the rotatory run is shown. For the first minute -a right rotation of the stripes- it is shown in cyan. The second minute, with left rotation of the stripes is color coded in magenta. Cyan is again marking right rotation of the stripes during the third minute. There is no significant correlation between walking behavior and stripe rotation.

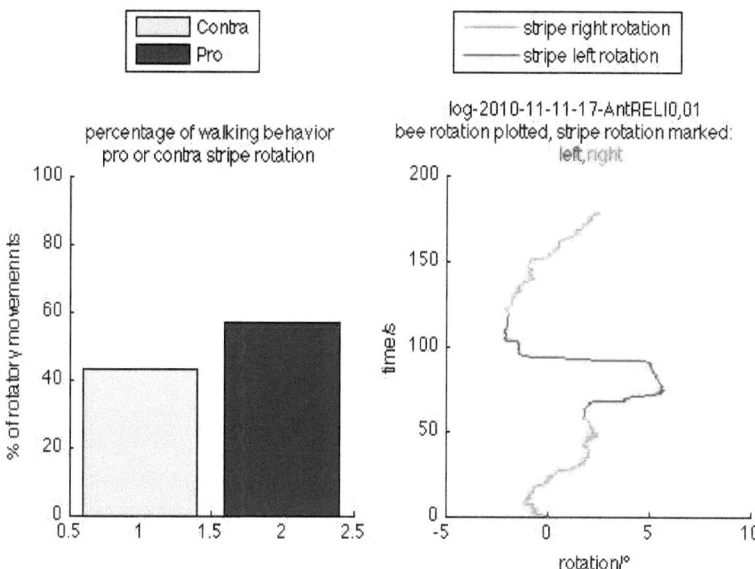

Fig.7: On the left, the percentage of walking behavior pro or contra stripe rotation is plotted. This means, not total time but single right or left rotation events, measured as difference between two consecutive mouse update events (see methods). Units on the x-axis are not meaningful. On the right side, the rotatory run is shown. For the first minute - right rotation of the stripes - it is shown in cyan. The second minute, with left rotation of the stripes is color coded in magenta. Cyan is again marking right rotation of the stripes during the third minute. The figure is the same as no 2 but with a threshold of 0.01 for rotatory data (see methods) in the histogram.

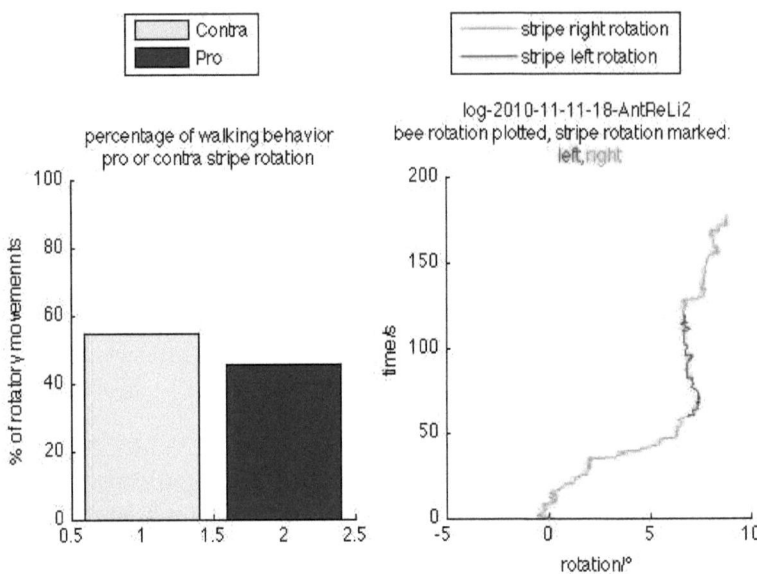

Fig.8: On the left, the percentage of walking behavior pro or contra stripe rotation is plotted. This means, not total time but single right or left rotation events, measured as difference between two consecutive mouse update events (see methods). Units on the x-axis are not meaningful. On the right side, the rotatory run is shown. For the first minute - right rotation of the stripes - it is shown in cyan. The second minute, with left rotation of the stripes is color coded in magenta. Cyan is again marking right rotation of the stripes during the third minute. There is no significant correlation between walking behavior and stripe rotation.

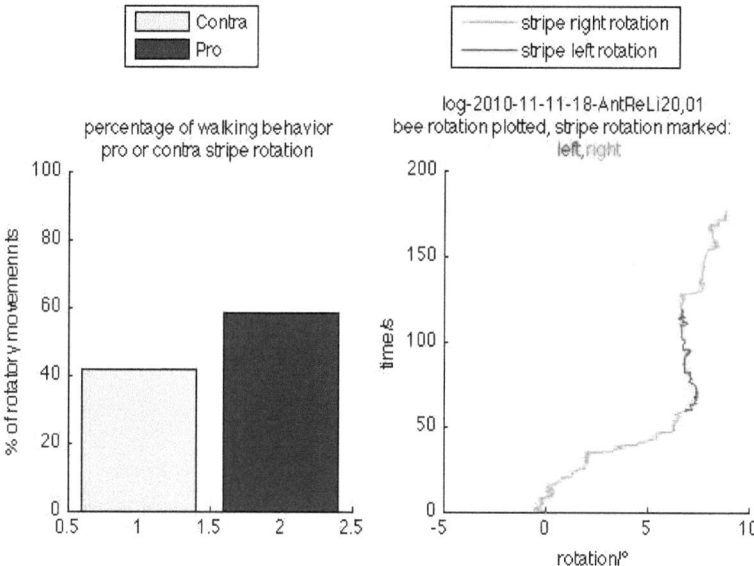

Fig.9: On the left, the percentage of walking behavior pro or contra stripe rotation is plotted. This means, not total time but single right or left rotation events, measured as difference between two consecutive mouse update events (see methods). Units on the x-axis are not meaningful. On the right side, the rotatory run is shown. For the first minute - right rotation of the stripes - it is shown in cyan. The second minute, with left rotation of the stripes is color coded in magenta. Cyan is again marking right rotation of the stripes during the third minute. The figure is the same as no 2 but with a threshold of 0.01 for rotatory data (see methods) in the histogram.

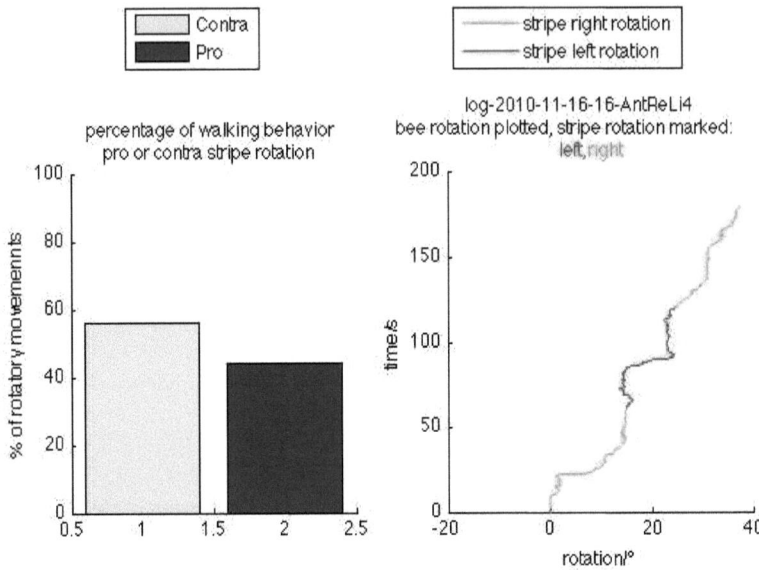

Fig.10: On the left, the percentage of walking behavior pro or contra stripe rotation is plotted. This means, not total time but single right or left rotation events, measured as difference between two consecutive mouse update events (see methods). Units on the x-axis are not meaningful. On the right side, the rotatory run is shown. For the first minute - right rotation of the stripes - it is shown in cyan. The second minute, with left rotation of the stripes is color coded in magenta. Cyan is again marking right rotation of the stripes during the third minute. There is no significant correlation between walking behavior and stripe rotation.

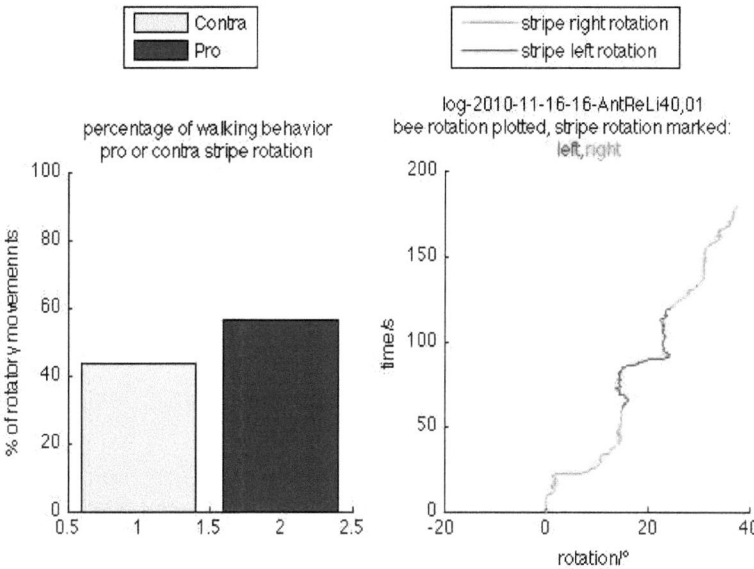

Fig.11: On the left, the percentage of walking behavior pro or contra stripe rotation is plotted. This means, not total time but single right or left rotation events, measured as difference between two consecutive mouse update events (see methods). Units on the x-axis are not meaningful. On the right side, the rotatory run is shown. For the first minute – right rotation of the stripes - it is shown in cyan. The second minute, with left rotation of the stripes is color coded in magenta. Cyan is again marking right rotation of the stripes during the third minute. The figure is the same as no 2 but with a threshold of 0.01 for rotatory data (see methods) in the histogram.

Clockwise stripe rotation experiments:

In these experiments the gratings are moving for three minutes in clockwise direction, therefore the rotatory run is shown in cyan. The histogram is the same as for the experiments above. An additional histogram is shown for a threshold of 0,01 degree for rotation values to eliminate possible tiny self rotations by the floating ball system itself. In addition the trajectory (including translational movements) is shown for every experiment. Neither for ants nor for bees significant optomotor walking responses are measured.

Bees:

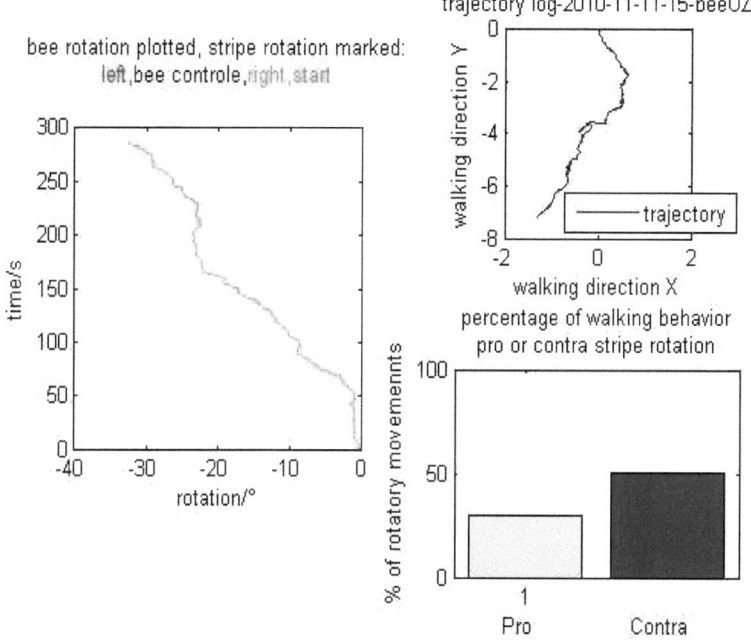

Fig.12: On the left, rotatory run is color coded in cyan, because the stripes are running in right direction for the whole three minutes of experimental time. Rotation is shown in degree on the x-axes, the time on the y-axes is shown in seconds. The upper right picture is showing the trajectory for the 3 minutes of experiment. Down right the percentage of walking behavior pro or contra stripe rotation is plotted. This means, not total time but single right or left rotation events, measured as difference between two consecutive mouse update events (see methods). The units on the x-axis are not meaningful.

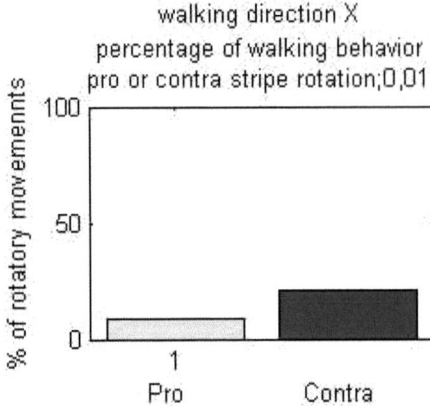

Fig.13: Percentage of walking behavior pro or contra stripe rotation is plotted for the bee log-2010-11-11-15-beeUZ. This means, not total time but single right or left rotation events, measured as difference between two consecutive mouse update events (see methods). Units on the x-axis are not meaningful. The figure is the same as the histogram in fig.12 but with a threshold of 0.01 for rotatory data (see methods) in the histogram.

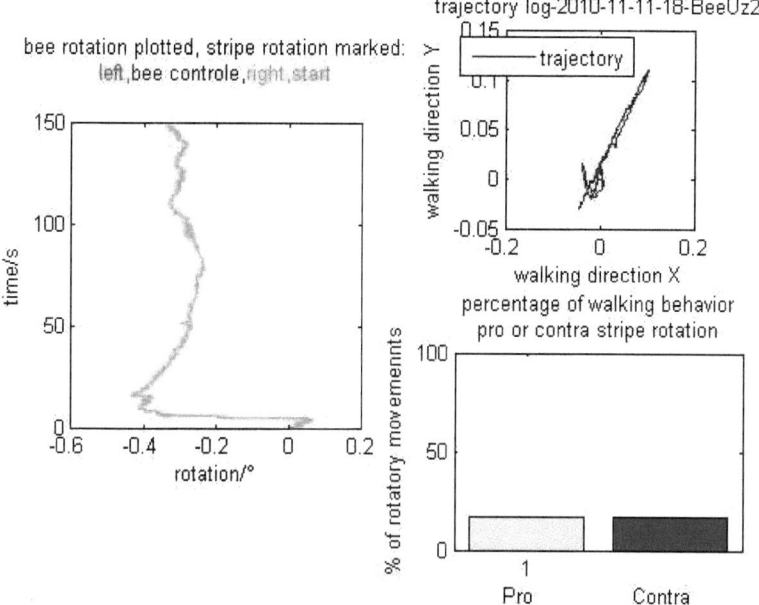

Fig.14: On the left, rotatory run is color coded in cyan, because the stripes are running in right direction for the whole three minutes of experimental time. Rotation is shown in degree on the x-axes, the time on the y-axes is shown in seconds. The upper right picture is showing the trajectory for the 3 minutes of experiment. Down right the percentage of walking behavior pro or contra stripe rotation is plotted. This means, not total time but single right or left rotation events, measured as difference between two consecutive mouse update events (see methods). The units on the x-axis are not meaningful.

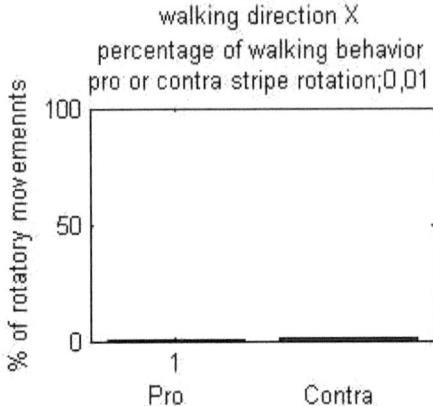

Fig.15: Percentage of walking behavior pro or contra stripe rotation is plotted for the bee log-2010-11-11-18-beeUZ2. This means, not total time but single right or left rotation events, measured as difference between two consecutive mouse update events (see methods). Units on the x-axis are not meaningful. The figure is the same as the histogram in fig.14 but with a threshold of 0.01 for rotatory data (see methods) in the histogram.

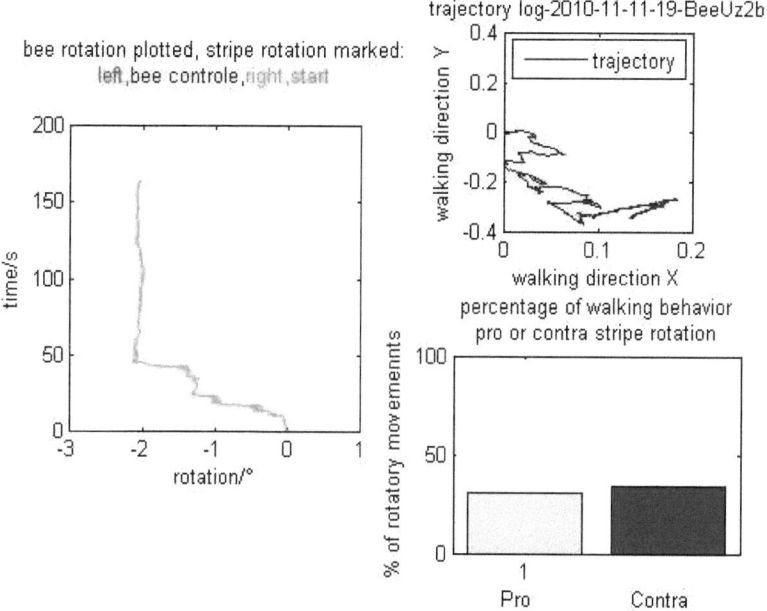

Fig.16: On the left, rotatory run is color coded in cyan, because the stripes are rotated in clockwise direction for the three minutes of experimental time. Rotation is shown in degree on the x-axes, the time on the y-axes is shown in seconds. The upper right picture is showing the trajectory for the 3 minutes of experiment. Down right the percentage of walking behavior pro or contra stripe rotation is plotted. This means, not total time but single right or left rotation events, measured as difference between two consecutive mouse update events (see methods). The units on the x-axis are not meaningful.

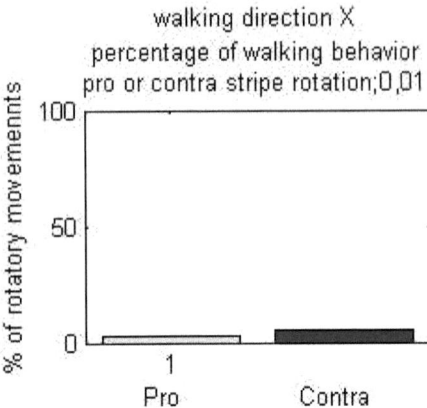

Fig.17: Percentage of walking behavior pro or contra stripe rotation is plotted for the bee log-2010-11-11-19-beeUZ2b. This means, not total time but single right or left rotation events, measured as difference between two consecutive mouse update events (see methods). Units on the x-axis are not meaningful. The figure is the same as the histogram in fig.16 but with a threshold of 0.01 for rotatory data (see methods) in the histogram.

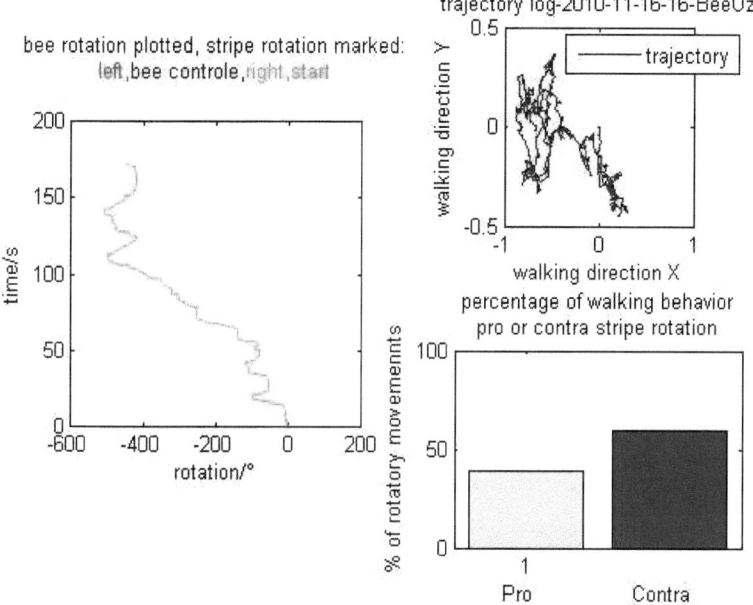

Fig.18: On the left, rotatory run is color coded in cyan, because the stripes are running in right direction for the whole three minutes of experimental time. Rotation is shown in degree on the x-axes, the time on the y-axes is shown in seconds. The upper right picture is showing the trajectory for the 3 minutes of experiment. Down right the percentage of walking behavior pro or contra stripe rotation is plotted. This means, not total time but single right or left rotation events, measured as difference between two consecutive mouse update events (see methods). The units on the x-axis are not meaningful.

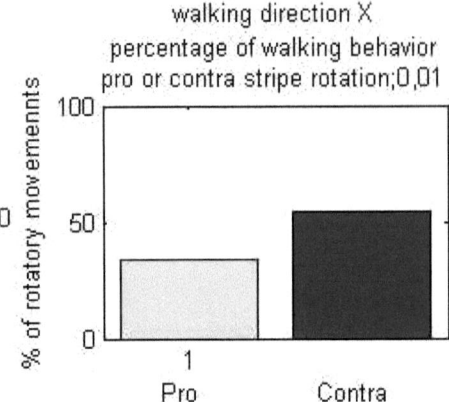

Fig.19: Percentage of walking behavior pro or contra stripe rotation is plotted for the bee log-2010-11-16-16-beeUZ. This means, not total time but single right or left rotation events, measured as difference between two consecutive mouse update events (see methods). Units on the x-axis are not meaningful. The histogram is the same as the one in fig.18 but with a threshold of 0.01 for rotatory data (see methods).

Ants:

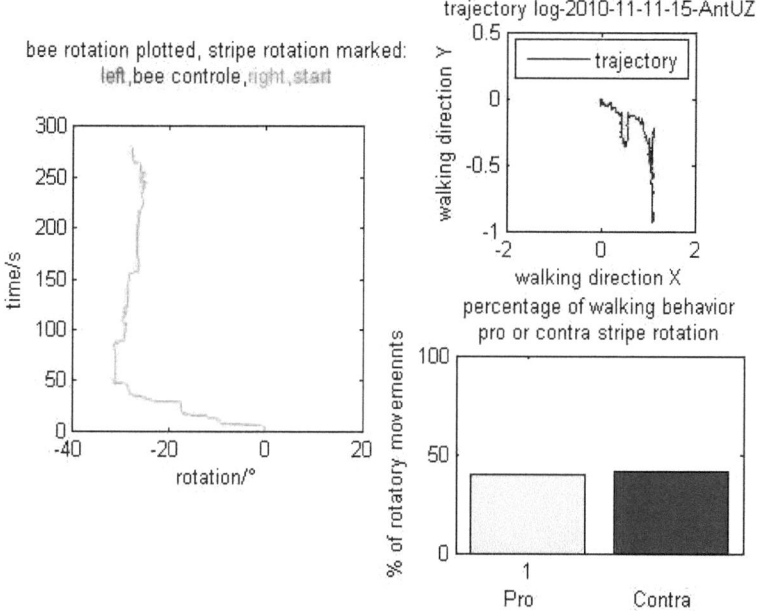

Fig.20: On the left, rotatory run is color coded in cyan, because the stripes are running in right direction for the whole three minutes of experimental time. Rotation is shown in degree on the x-axes, the time on the y-axes is shown in seconds. The upper right picture is showing the trajectory for the 3 minutes of experiment. Down right the percentage of walking behavior pro or contra stripe rotation is plotted. This means, not total time but single right or left rotation events, measured as difference between two consecutive mouse update events (see methods). The units on the x-axis are not meaningful.

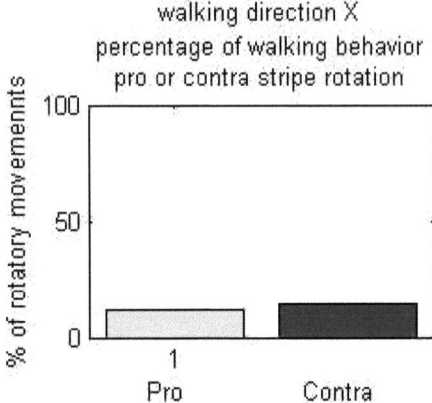

Fig.21: Percentage of walking behavior pro or contra stripe rotation is plotted for the ant log-2010-11-11-15-antUZ. This means, not total time but single right or left rotation events, measured as difference between two consecutive mouse update events (see methods). Units on the x-axis are not meaningful. The histogram is the same as the one in fig. 20 but with a threshold of 0.01 for rotatory data (see methods).

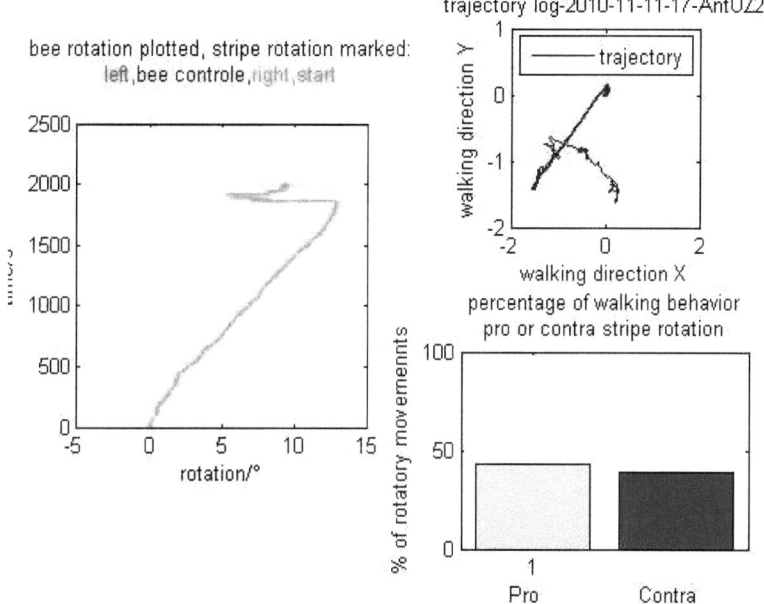

Fig.22: On the left, rotatory run is color coded in cyan, because the stripes are running in right direction for the whole three minutes of experimental time. Rotation is shown in degree on the x-axes, the time on the y-axes is shown in seconds. The upper right picture is showing the trajectory for the 3 minutes of experiment. Down right the percentage of walking behavior pro or contra stripe rotation is plotted. This means, not total time but single right or left rotation events, measured as difference between two consecutive mouse update events (see methods). The units on the x-axis are not meaningful.

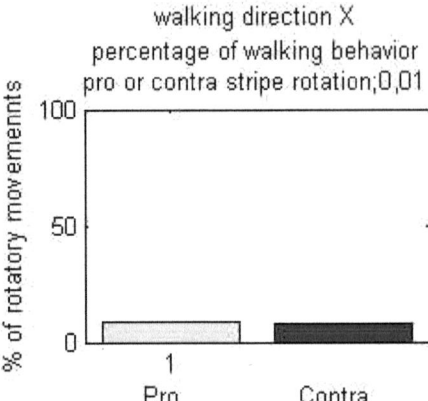

Fig.23: Percentage of walking behavior pro or contra stripe rotation is plotted for the ant log-2010-11-11-17-antUZ2. This means, not total time but single right or left rotation events, measured as difference between two consecutive mouse update events (see methods). Units on the x-axis are not meaningful. The histogram is the same as the one in fig.22 but with a threshold of 0.01 for rotatory data (see methods).

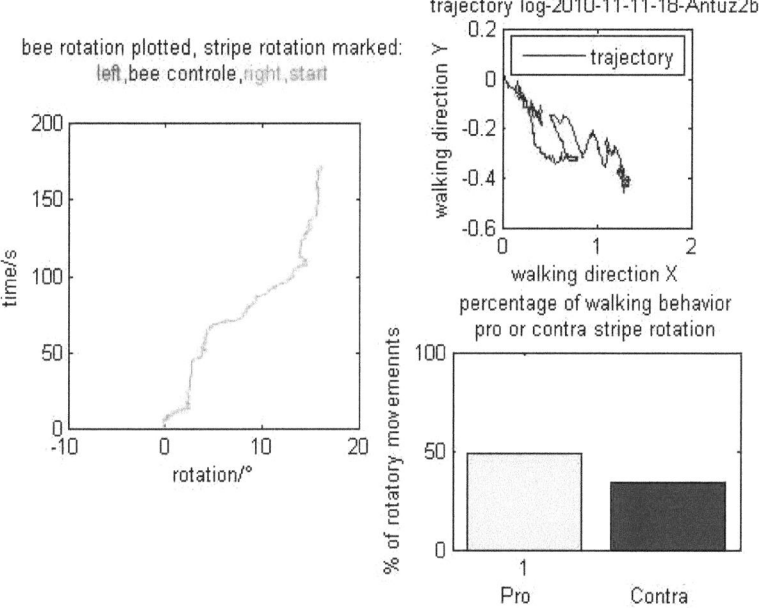

Fig.24: On the left, rotatory run is color coded in cyan, because the stripes are running in right direction for the whole three minutes of experimental time. Rotation is shown in degree on the x-axes, the time on the y-axes is shown in seconds. The upper right picture is showing the trajectory for the 3 minutes of experiment. Down right the percentage of walking behavior pro or contra stripe rotation is plotted. This means, not total time but single right or left rotation events, measured as difference between two consecutive mouse update events (see methods). The units on the x-axis are not meaningful.

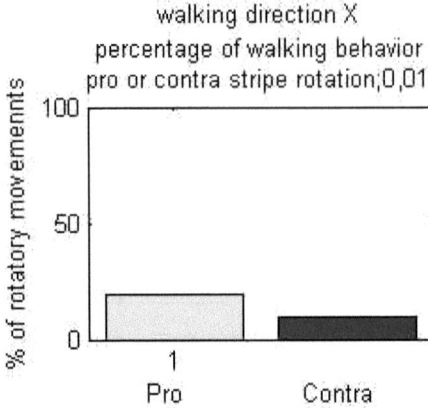

Fig.25: Percentage of walking behavior pro or contra stripe rotation is plotted for the ant log-2010-11-11-18-antUZ2b. This means, not total time but single right or left rotation events, measured as difference between two consecutive mouse update events (see methods). Units on the x-axis are not meaningful. The histogram is the same as the one in fig.24 but with a threshold of 0.01 for rotatory data (see methods).

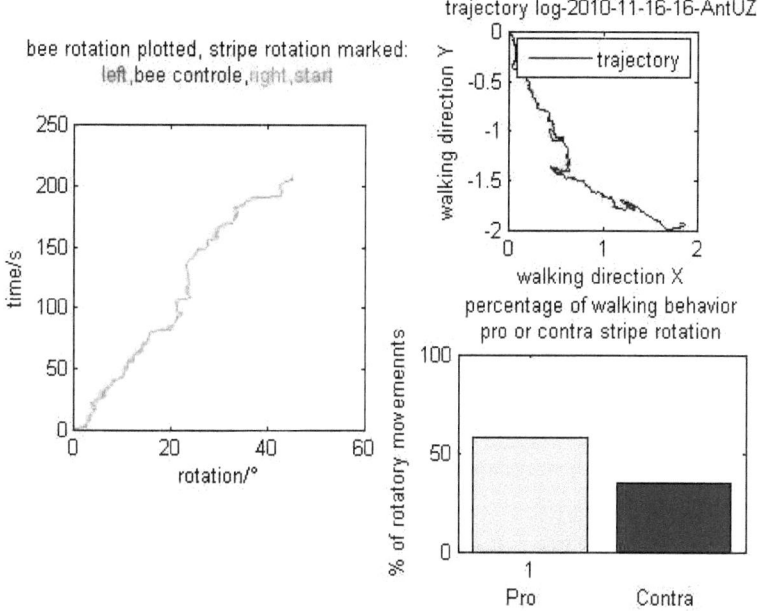

Fig.26: On the left, rotatory run is color coded in cyan, because the stripes are running in right direction for the whole three minutes of experimental time. Rotation is shown in degree on the x-axes, the time on the y-axes is shown in seconds. The upper right picture is showing the trajectory for the 3 minutes of experiment. Down right the percentage of walking behavior pro or contra stripe rotation is plotted. This means, not total time but single right or left rotation events, measured as difference between two consecutive mouse update events (see methods). The units on the x-axis are not meaningful.

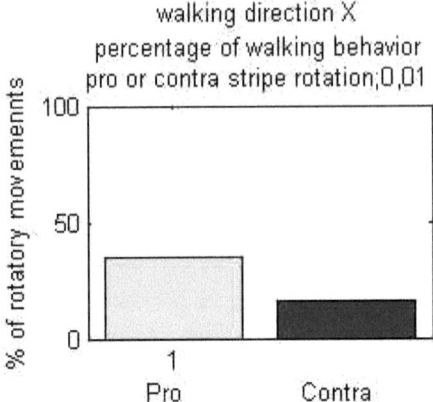

Fig.27: Percentage of walking behavior pro or contra stripe rotation is plotted for the ant log-2010-11-16-16-antUZ. This means, not total time but single right or left rotation events, measured as difference between two consecutive mouse update events (see methods). Units on the x-axis are not meaningful. The histogram is the same as the one in fig.26 but with a threshold of 0.01 for rotatory data (see methods).

Counterclockwise stripe rotation experiments:

The same experiment as shown above but the stripes are running in counterclockwise direction for three minutes. Therefore the rotatory run is color coded in magenta. As above no significant optomotor responses are detected.

Bees:

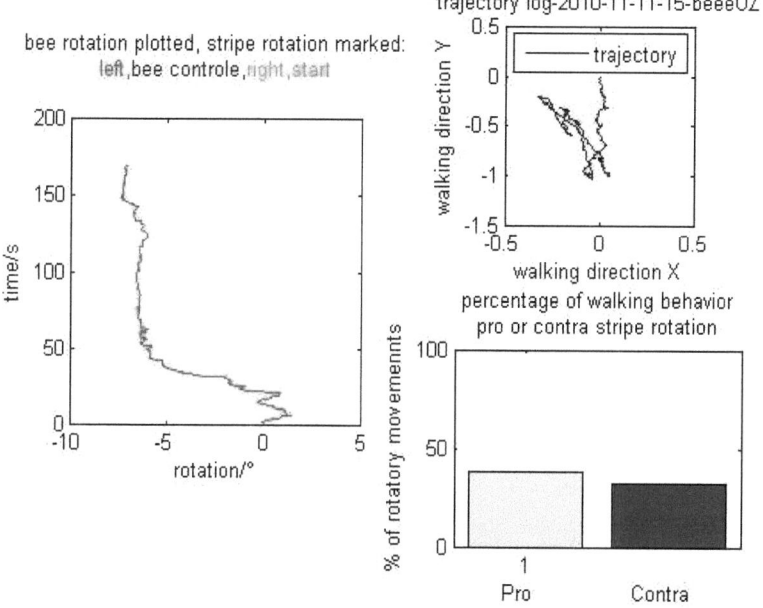

Fig.28: On the left, rotatory run is color coded in magenta, because the stripes are running in left direction for the whole three minutes of experimental time. Rotation is shown in degree on the x-axes, the time on the y-axes is shown in seconds. The upper right picture is showing the trajectory for the 3 minutes of experiment. Down right the percentage of walking behavior pro or contra stripe rotation is plotted. This means, not total time but single right or left rotation events, measured as difference between two consecutive mouse update events (see methods). The units on the x-axis are not meaningful.

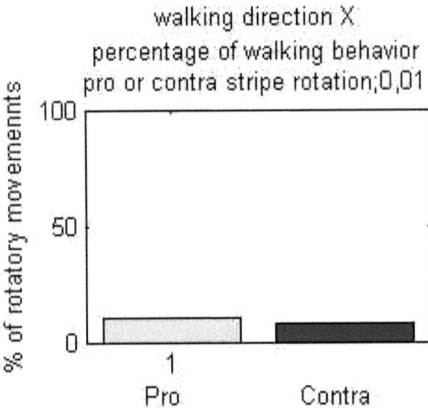

Fig.29: Percentage of walking behaviour pro or contra stripe rotation is plotted for the bee log-2010-11-11-15-beeeUZ. This means, not total time but single right or left rotation events, measured as difference between two consecutive mouse update events (see methods). Units on the x-axis are not meaningful. The histogram is the same as the one in fig.28 but with a threshold of 0.01 for rotatory data (see methods).

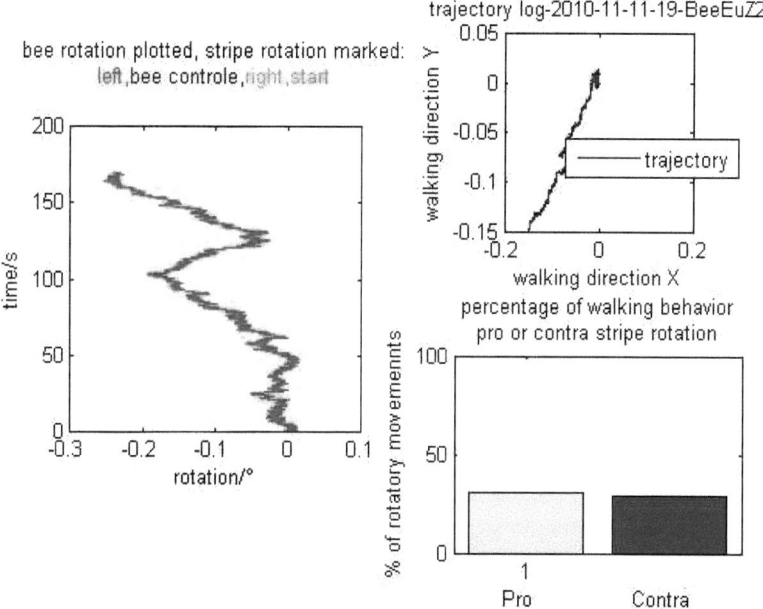

Fig.30: On the left, rotatory run is color coded in magenta, because the stripes are running in left direction for the whole three minutes of experimental time. Rotation is shown in degree on the x-axes, the time on the y-axes is shown in seconds. The upper right picture is showing the trajectory for the 3 minutes of experiment. Down right the percentage of walking behavior pro or contra stripe rotation is plotted. This means, not total time but single right or left rotation events, measured as difference between two consecutive mouse update events (see methods). The units on the x-axis are not meaningful.

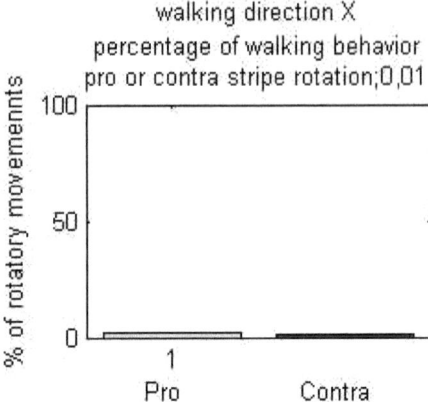

Fig.31: Percentage of walking behavior pro or contra stripe rotation is plotted for the bee log-2010-11-11-19-beeEUZ2. This means, not total time but single right or left rotation events, measured as difference between two consecutive mouse update events (see methods). Units on the x-axis are not meaningful. The histogram is the same as the one in fig.30 but with a threshold of 0.01 for rotatory data (see methods).

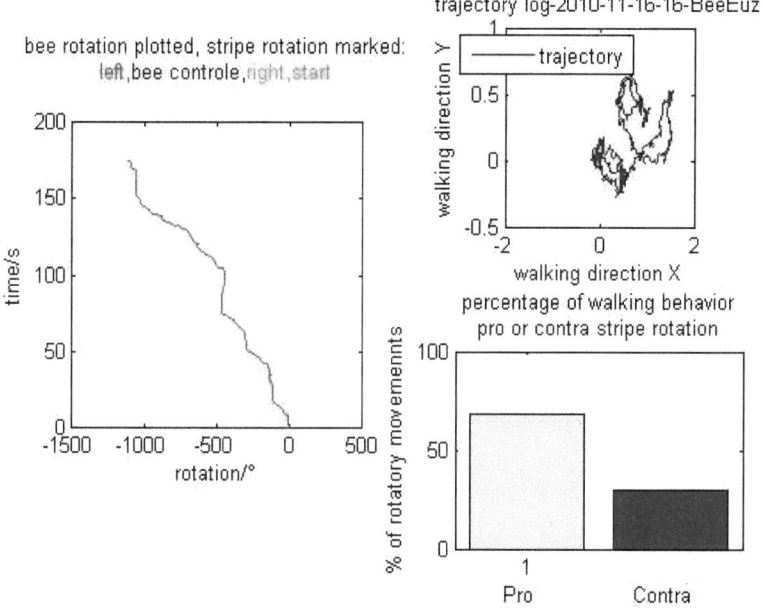

Fig.32: On the left, rotatory run is color coded in magenta, because the stripes are running in left direction for the whole three minutes of experimental time. Rotation is shown in degree on the x-axes, the time on the y-axes is shown in seconds. The upper right picture is showing the trajectory for the 3 minutes of experiment. Down right the percentage of walking behavior pro or contra stripe rotation is plotted. This means, not total time but single right or left rotation events, measured as difference between two consecutive mouse update events (see methods). The units on the x-axis are not meaningful.

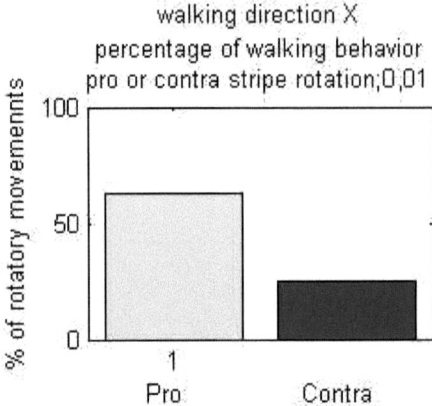

Fig.33: Percentage of walking behavior pro or contra stripe rotation is plotted for the bee log-2010-11-16-16-beeEUZ. This means, not total time but single right or left rotation events, measured as difference between two consecutive mouse update events (see methods). Units on the x-axis are not meaningful. The histogram is the same as the one in fig.32 but with a threshold of 0.01 for rotatory data (see methods).

Ants:

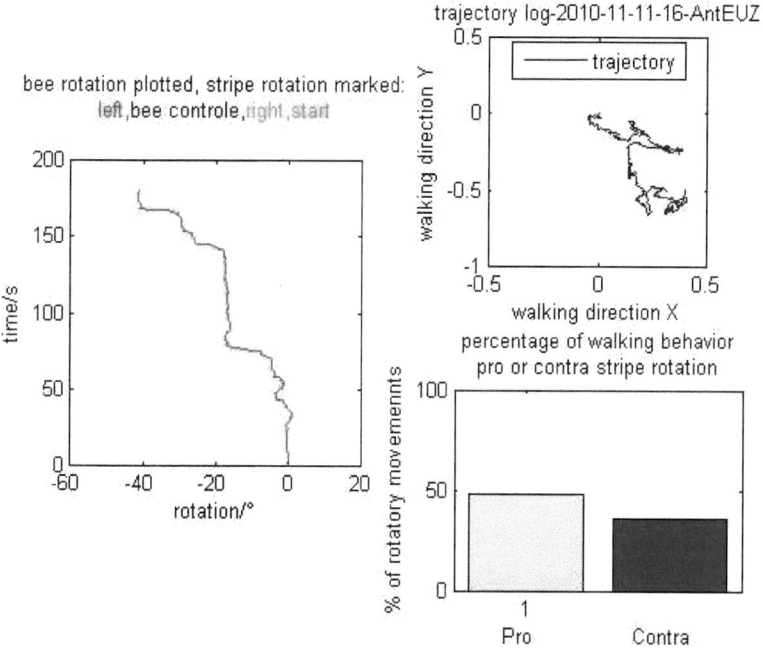

Fig.34: On the left, rotatory run is color coded in magenta, because the stripes are running in left direction for the whole three minutes of experimental time. Rotation is shown in degree on the x-axes, the time on the y-axes is shown in seconds. The upper right picture is showing the trajectory for the 3 minutes of experiment. Down right the percentage of walking behavior pro or contra stripe rotation is plotted. This means, not total time but single right or left rotation events, measured as difference between two consecutive mouse update events (see methods). The units on the x-axis are not meaningful.

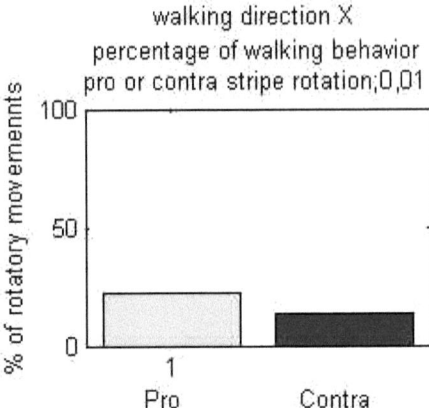

Fig.35: Percentage of walking behavior pro or contra stripe rotation is plotted for the ant log-2010-11-11-16-antEUZ. This means, not total time but single right or left rotation events, measured as difference between two consecutive mouse update events (see methods). Units on the x-axis are not meaningful. The histogram is the same as the one in fig.34 but with a threshold of 0.01 for rotatory data (see methods).

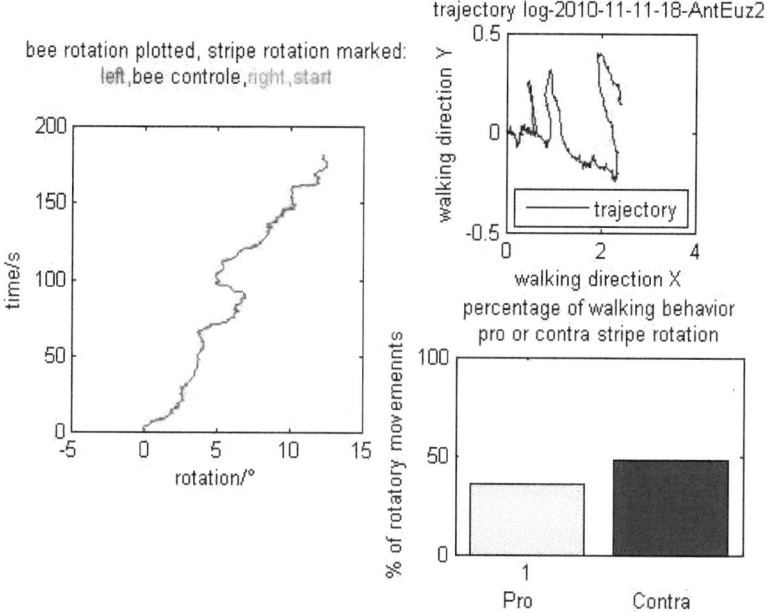

Fig.36: On the left, rotatory run is color coded in magenta, because the stripes are running in left direction for the whole three minutes of experimental time. Rotation is shown in degree on the x-axes, the time on the y-axes is shown in seconds. The upper right picture is showing the trajectory for the 3 minutes of experiment. Down right the percentage of walking behavior pro or contra stripe rotation is plotted. This means, not total time but single right or left rotation events, measured as difference between two consecutive mouse update events (see methods). The units on the x-axis are not meaningful.

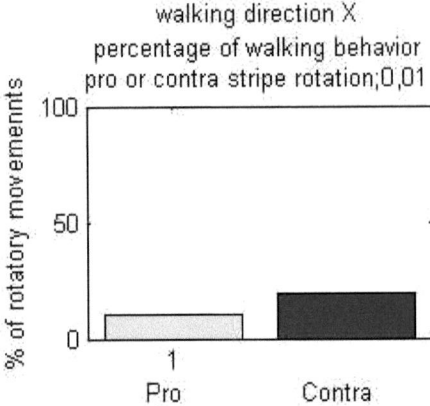

Fig.37: Percentage of walking behavior pro or contra stripe rotation is plotted for the ant log-2010-11-11-18-antEUZ2. This means, not total time but single right or left rotation events, measured as difference between two consecutive mouse update events (see methods). Units on the x-axis are not meaningful. The histogram is the same as the one in fig.36 but with a threshold of 0.01 for rotatory data (see methods).

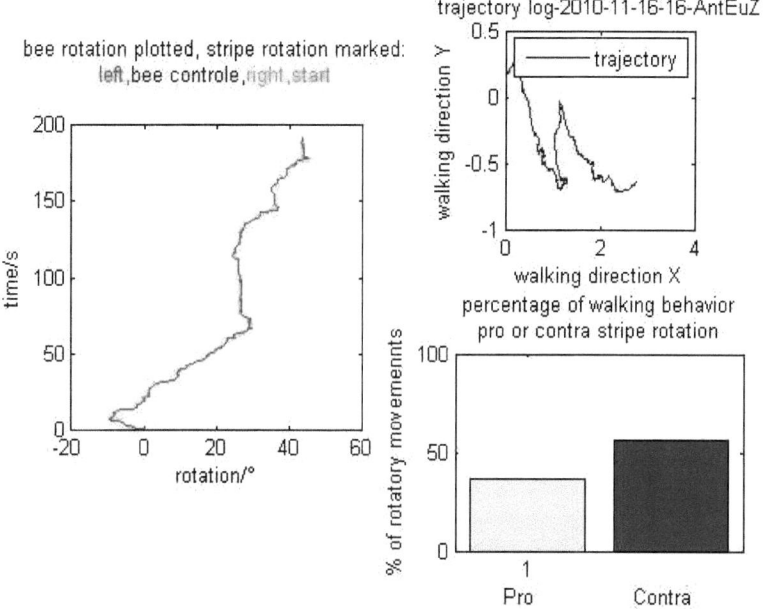

Fig.38: On the left, rotatory run is color coded in magenta, because the stripes are running in left direction for the whole three minutes of experimental time. Rotation is shown in degree on the x-axes, the time on the y-axes is shown in seconds. The upper right picture is showing the trajectory for the 3 minutes of experiment. Down right the percentage of walking behavior pro or contra stripe rotation is plotted. This means, not total time but single right or left rotation events, measured as difference between two consecutive mouse update events (see methods). The units on the x-axis are not meaningful.

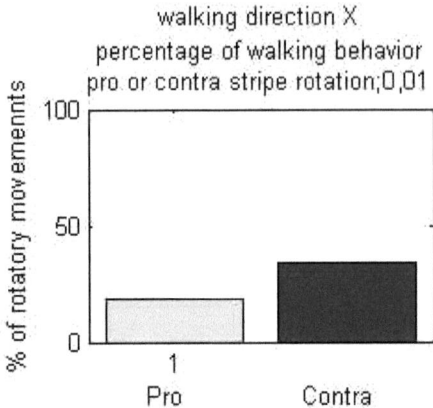

Fig.39: Percentage of walking behavior pro or contra stripe rotation is plotted for the ant log-2010-11-16-16-antEUZ. This means, not total time but single right or left rotation events, measured as difference between two consecutive mouse update events (see methods). Units on the x-axis are not meaningful. The histogram is the same as the one in fig.38 but with a threshold of 0.01 for rotatory data (see methods).

Control experiments:

For these experiments stripes were presented without movement for three minutes. Therefore the direction is not called Pro or Contra anymore but right and left. There is no difference in animal rotation behavior in comparison to the above mentioned expermiments with stripe rotation.

Bees:

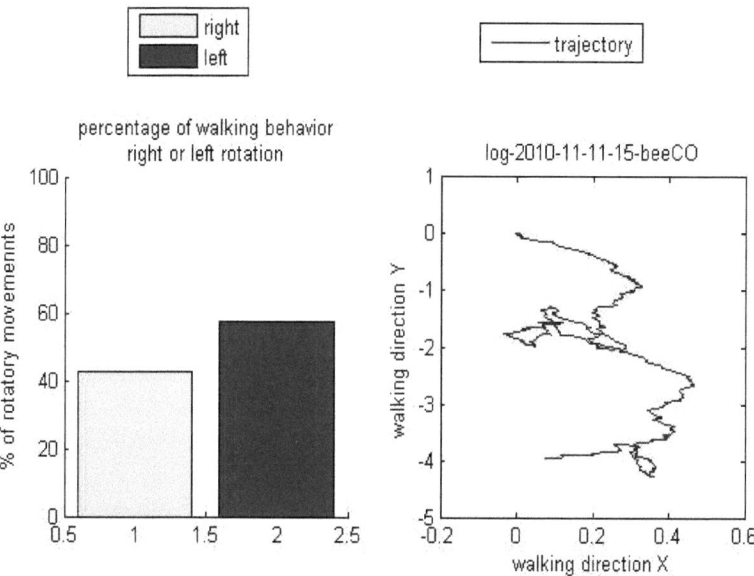

Fig.40: The histogram on the left is showing the percentage of right or left rotation events by the animal onto the floating ball (see methods), detected as difference of rotation values for two consecutive mouse update events. There was no stripe rotation at all, but the same environment as in the test situations for a period of three minutes. The trajectory is shown on the right side.

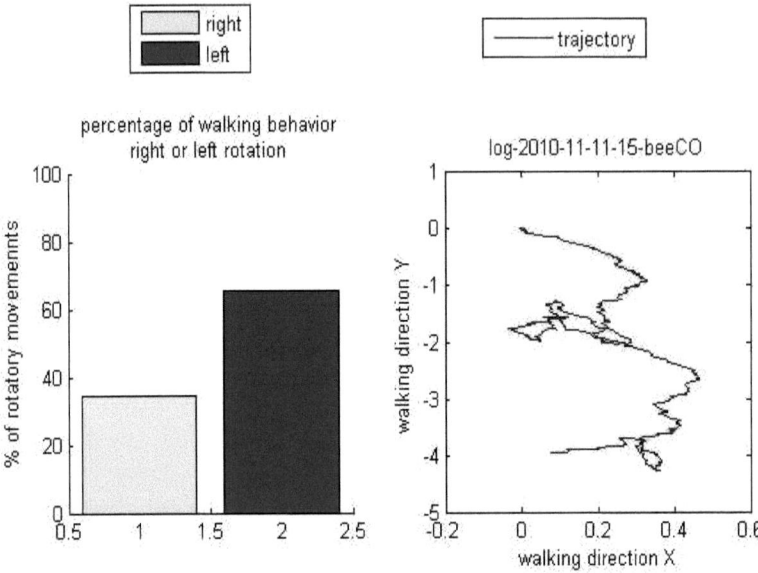

Fig.41: The histogram on the left is showing the percentage of right or left rotation events by the animal onto the floating ball (see methods), detected as difference of rotation values (threshold 0.01) for two consecutive mouse update events. There was no stripe rotation at all, but the same environment as in the test situations for a period of three minutes. The trajectory is shown on the right side.

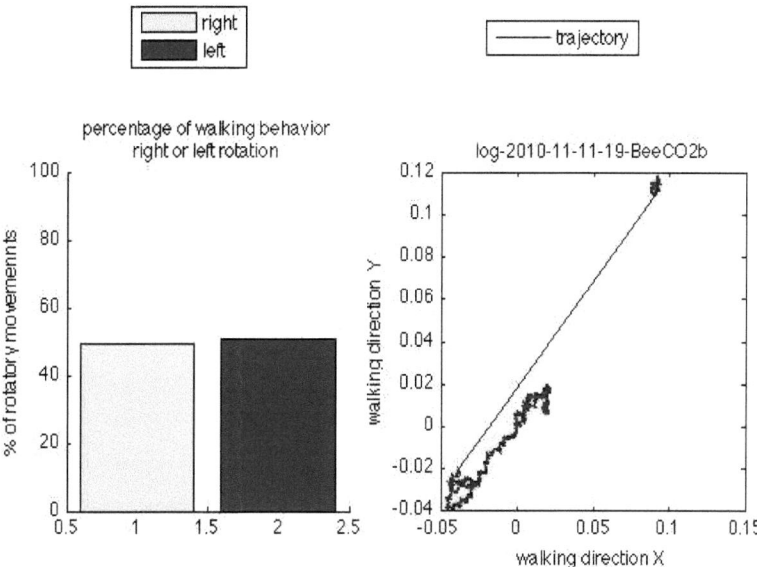

Fig.42: The histogram on the left is showing the percentage of right or left rotation events by the animal onto the floating ball (see methods), detected as difference of rotation values for two consecutive mouse update events. There was no stripe rotation at all, but the same environment as in the test situations for a period of three minutes. The trajectory is shown on the right side.

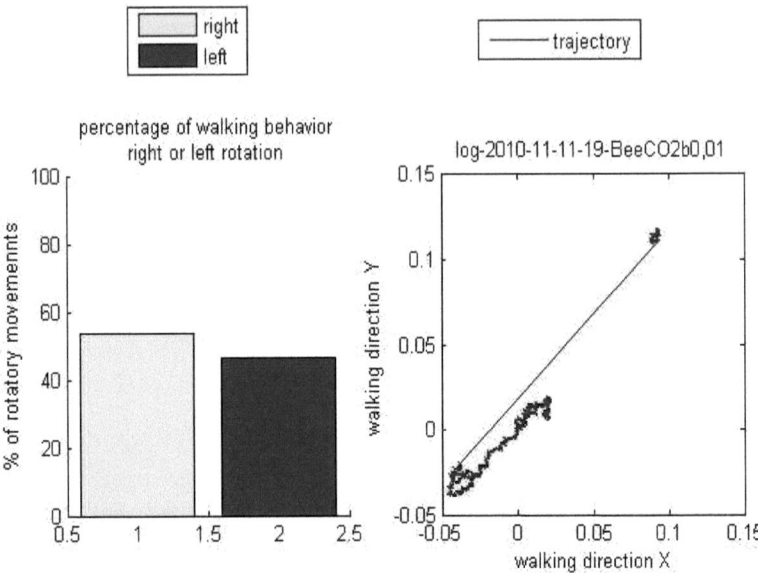

Fig.43: The histogram on the left is showing the percentage of right or left rotation events by the animal onto the floating ball (see methods), detected as difference of rotation values (threshold 0.01) for two consecutive mouse update events. There was no stripe rotation at all, but the same environment as in the test situations for a period of three minutes. The trajectory is shown on the right side.

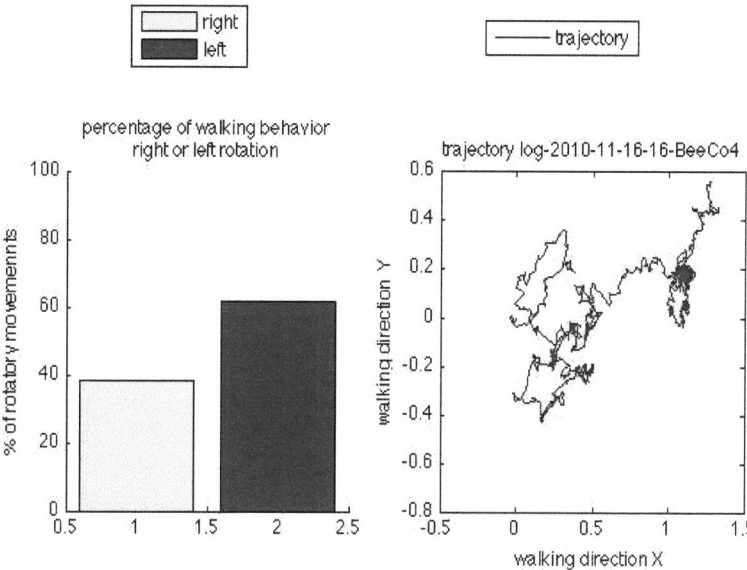

Fig.44: The histogram on the left is showing the percentage of right or left rotation events by the animal onto the floating ball (see methods), detected as difference of rotation values for two consecutive mouse update events. There was no stripe rotation at all, but the same environment as in the test situations for a period of three minutes. The trajectory is shown on the right side.

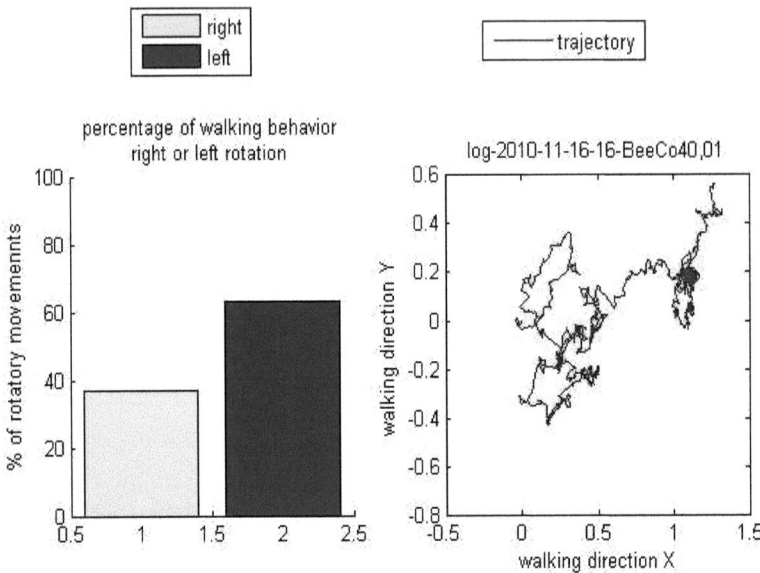

Fig.45: The histogram on the left is showing the percentage of right or left rotation events by the animal onto the floating ball (see methods), detected as difference of rotation values (with threshold 0.01) for two consecutive mouse update events. There was no stripe rotation at all, but the same environment as in the test situations for a period of three minutes. The trajectory is shown on the right side.

Ants:

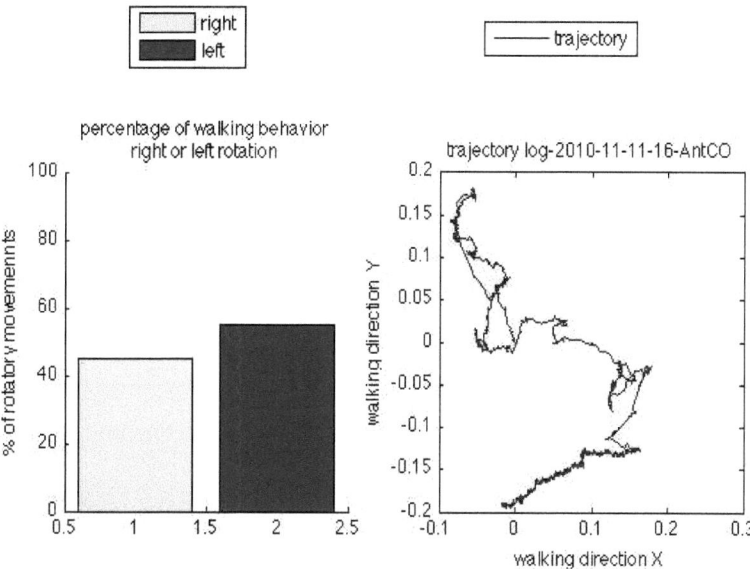

Fig.46: The histogram on the left is showing the percentage of right or left rotation events by the animal onto the floating ball (see methods), detected as difference of rotation values for two consecutive mouse update events. There was no stripe rotation at all, but the same environment as in the test situations for a period of three minutes. The trajectory is shown on the right side.

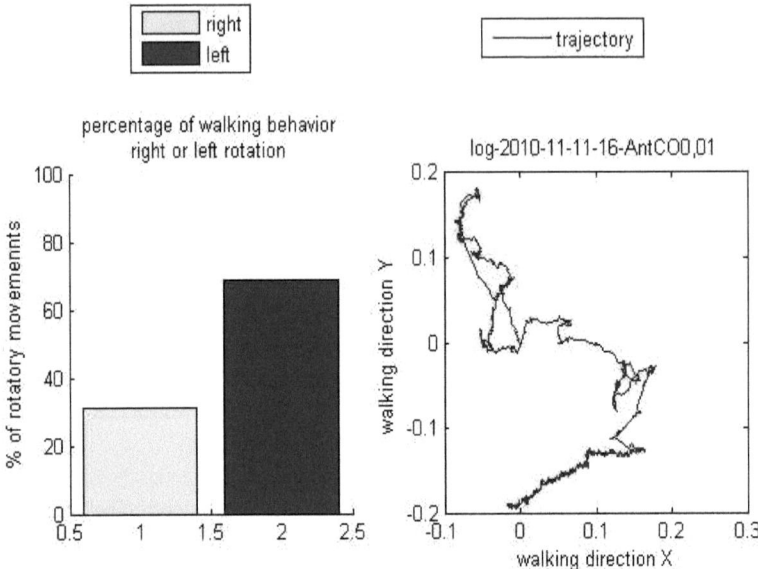

Fig.47: The histogram on the left is showing the percentage of right or left rotation events by the animal onto the floating ball (see methods), detected as difference of rotation values (with threshold 0.01) for two consecutive mouse update events. There was no stripe rotation at all, but the same environment as in the test situations for a period of three minutes. The trajectory is shown on the right side.

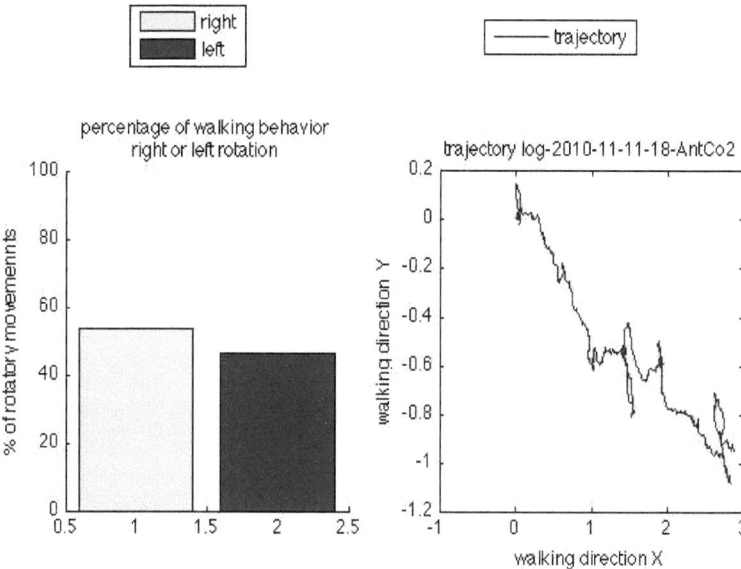

Fig.48: The histogram on the left is showing the percentage of right or left rotation events by the animal onto the floating ball (see methods), detected as difference of rotation values for two consecutive mouse update events. There was no stripe rotation at all, but the same environment as in the test situations for a period of three minutes. The trajectory is shown on the right side.

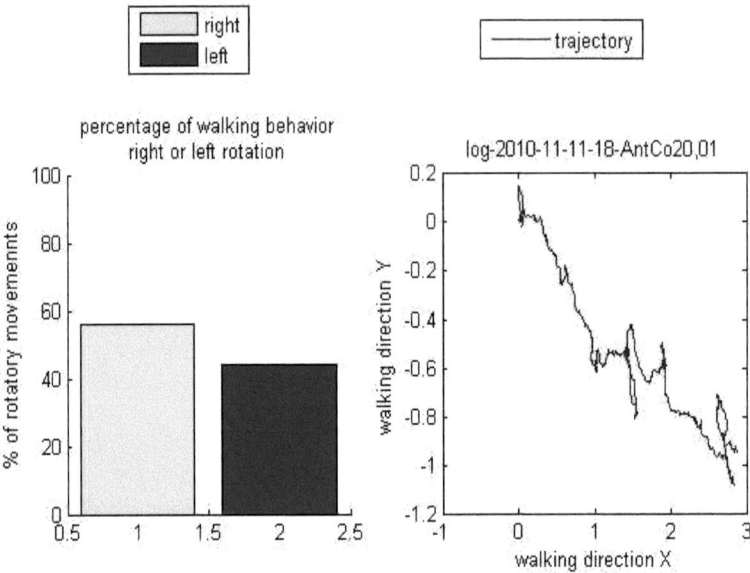

Fig.49: The histogram on the left is showing the percentage of right or left rotation events by the animal onto the floating ball (see methods), detected as difference of rotation values (with threshold 0.01) for two consecutive mouse update events. There was no stripe rotation at all, but the same environment as in the test situations for a period of three minutes. The trajectory is shown on the right side.

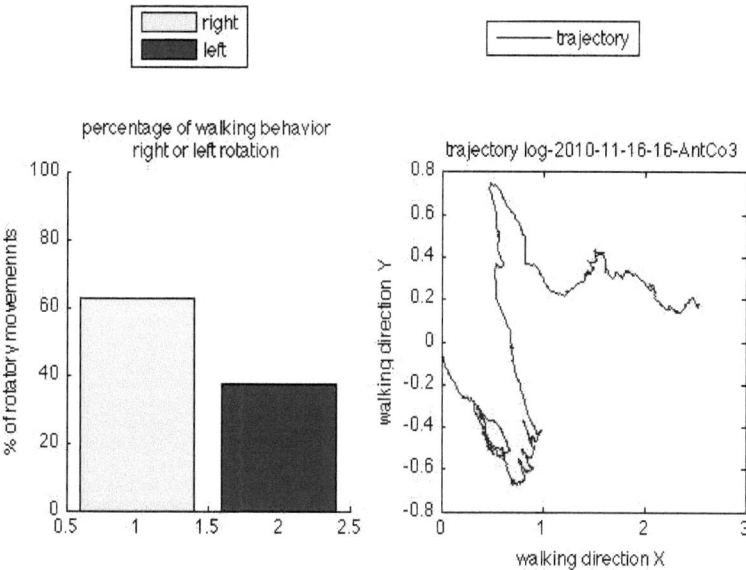

Fig.50: The histogram on the left is showing the percentage of right or left rotation events by the animal onto the floating ball (see methods), detected as difference of rotation values for two consecutive mouse update events. There was no stripe rotation at all, but the same environment as in the test situations for a period of three minutes. The trajectory is shown on the right side.

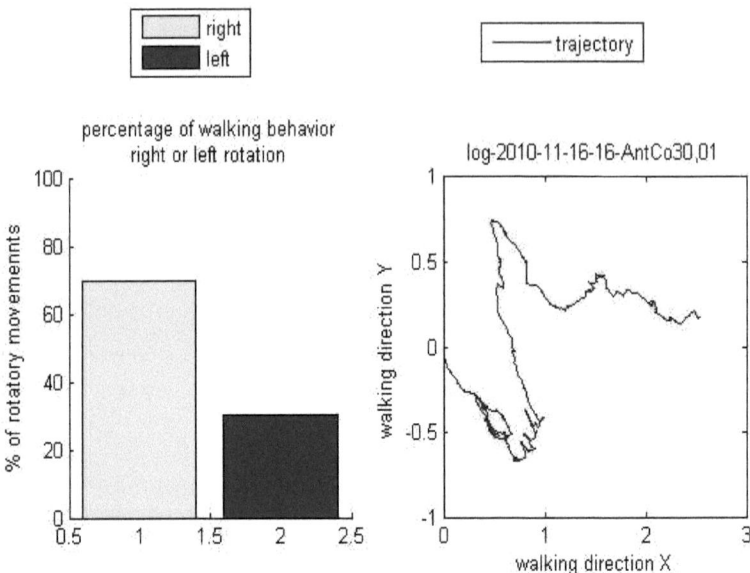

Fig.51: The histogram on the left is showing the percentage of right or left rotation events by the animal onto the floating ball (see methods), detected as difference of rotation values (with threshold 0.01) for two consecutive mouse update events. There was no stripe rotation at all, but the same environment as in the test situations for a period of three minutes. The trajectory is shown on the right side.

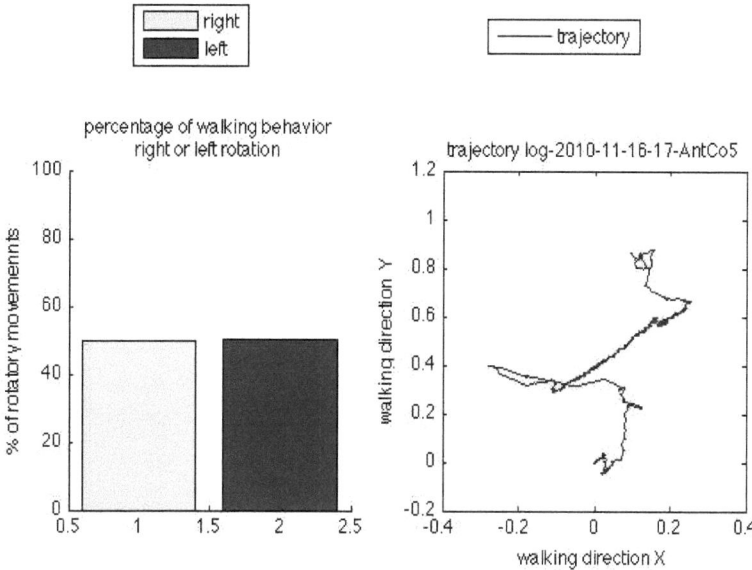

Fig.52: The histogram on the left is showing the percentage of right or left rotation events by the animal onto the floating ball (see methods), detected as difference of rotation values for two consecutive mouse update events. There was no stripe rotation at all, but the same environment as in the test situations for a period of three minutes. The trajectory is shown on the right side.

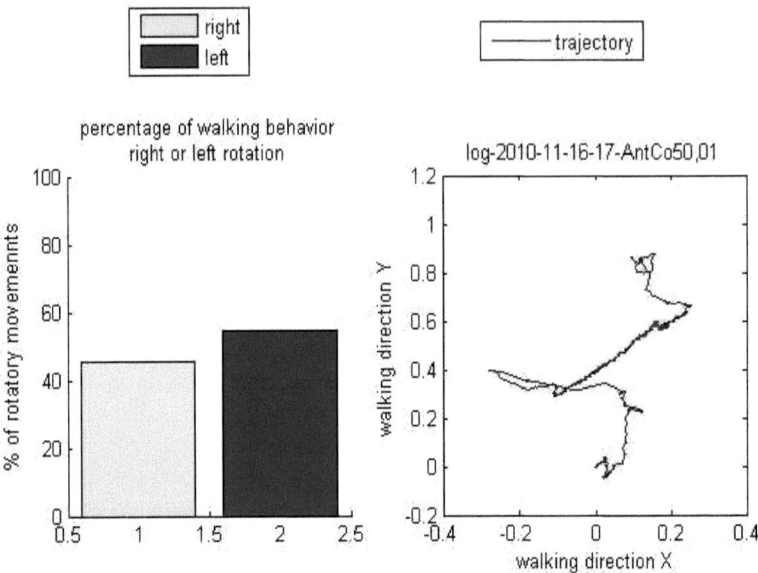

Fig.53: The histogram on the left is showing the percentage of right or left rotation events by the animal onto the floating ball (see methods), detected as difference of rotation values (with threshold 0.01) for two consecutive mouse update events. There was no stripe rotation at all, but the same environment as in the test situations for a period of three minutes. The trajectory is shown on the right side.

Discussion. Neither stationary walking Bees nor Ants show optomotor walking behavior. With respect to the findings of Haag et al. (2010) the results of this study are not necessariliy surprising. Additional wind sensitive inputs have not been presented. Since the illumination potentiates optomotor behavior (Kunze, 1961), the illumination level inside the drum might have been too low, to elicit optomotor walking behavior. Furthermore, it can not be excluded, that optomotor head movements have taken place. These movements are very fast and hardly observed bye eye (Böddeker&Hemmi, 2010). In contrast to forager bees, drones show obvious optomotor head movements in a reliable manner (own observation). This difference in optomotor head movements between male and female bees might arise from the drone's specialized optic apparatus, which is mainly constructed to detect the queen during flight. The high variability in walking behavior between animals might be explained by different gain states, as Rosner et al. (2009) have proposed for the blowfly. Brain derived gain states could also account for optomotor head movements in male but not female bees. It would be interesting to test younger bees for optomotor walking behavior. Young working bees with indoor tasks can learn a visual PER paradigm, without cutting the antennae (Dobrin et al., 2012) which implies a different kind of sensory integration. As Srinivasan et al. (1991) have shown, bees tend to loose optomotor behavior in an ongoing learning task if they have to pass a tunnel with stripes on one wall. Considering these experiments I would strongly agree their hypothesis that the optomotor behavior, which depends on contrast frequency (Kunze, 1961), is suppressed by or in competition to other visuomotor tasks. For example the centering which is calculated from angular speed.

References:

Böddeker, N., Hemmi, J.M. (2010) Visual gaze control during peering flight manoeuvres in honeybees. Proc. R. Soc. B. Biol. Sci. **277**:1209–1217.

Chow, D.M., Theobald, J.C., Frye, M.A. (2011) An Olfactory Circuit Increases the Fidelity of Visual Behavior. The Journal of Neuroscience **31(42)**: 15035–15047.

Dobrin, S.E., Fahrbach, S.E. (2012) Visual associative learning in restrained honey bees with intact antennae. PloS One **7(6)**:e37666. Epub 2012 Jun 6.

Haag et al. (2010). Central gating of fly optomotor response. PNAS **107**: 20104-20109.

Hung, Y.-S., van Kleef, J. P., Ibbotson, M. R. (2011) Visual response properties of neck motor neurons in the honeybee. J. Comp. Physiol. A **197**: 1173–1187.

Kunze, P. (1961) Untersuchung des bewegungssehens fixiert fliegender Bienen. Zeitschrift fur Vergleichende Physiologie **44**: 656–684.

Lehrer, M. (1996) Small-scale navigation in the honeybee: active acquisition of visual information about the goal. J. Exp. Biol. **199**: 253–261.

Lehrer, M. (1991) Bees which turn back and look. Naturwissenschaften **78**: 274–276.

Reichardt W. (1969) Movement perception in insects. In *Processing of optical data by organisms and by machines.* (Edited by Reichardt W.). Academic Press, New York.

Rosner etal. (2009). Variability of blowfly head optomotor responses. *The Journal of Experimental biology* **212**: 1170-1184.

Srinivasan, M. V., Zhang, S. (2004) Visual motor computations in insects. Annu. Rev. Neurosci. **27**: 679–696.

Srinivasan et al. (1991) Range perception through apparent image speed in freely flying honeybees. *Visual Neuroscience* **6**: 519-535.

Srinivasan M.V., Dvorak D. R. (1978) The waterfall illusion in an insect visual system. *Vision Research* **19**: 1435-1437.

Voss, R., Zeil, J. (1998) Active vision in insects: an analysis of object-directed zig-zag flights in wasps (Odynerus spinipes, Eumenidae). J. Comp. Physiol. A **182**: 377–387.

4. New methods for extracellular brain recordings in stationary and freely walking honeybees during decision making and navigation

Abstract. Are studies of complex behavioral actions and electrophysiology in small insects, like honeybees, mutually exclusive? In a first approach to study behavioral decision making in insects of about honeybee size, a method for extracellular brain recordings in a virtual environment, that simulates a simplified 2D world for honeybees walking stationary on an air-supported spherical treadmill that controls the environment, is decribed. The rotatory and translatory movements of the treadmill are translated in respective changes of the visual patterns. Honeybees respond to these patterns and show different walking trajectories in the virtual environment. During long lasting extracellular recordings the neural activity precedes turning manoeuvres and lasts longer than the motor activity indicating choice dependent neural activity.

In a second approach, micro-plugs for chronical implantation in the bee brain were developed. They are suitable to study extracellular brain electrophysiology and complex behavioral actions in freely walking honeybees in a real setting.

Introduction The search for neural correlates of behavioural decision making during navigation requires the combination of two procedures that are usually very difficult to combine, stable recording from neurons and free movement of the animal in a rather natural environment. Two approaches have been followed in the past, monitoring neural activity of animals (usually rats) navigating in a small space, and animals (or humans) navigating in a virtual reality environment (VE). The latter has the advantage that the simulated environment can be large and fully manipulated, but the disadvantages related to compromised sensory feedback provided by the moving visual world and the stationary conditions of the animal. Nevertheless animals and humans can navigate in a virtual reality set-up that produces the relevant visual feedback to the intended movements in 2D (design guidelines for human navigation in VE (Vinson, 1999) can be found at: http://cogprints.org/3297/). Multiple devices of this kind have been developed for human subjects exploring behavioural navigation, biofeedback and psychotherapy, and were successfully combined with neural recordings both of EEG and local field potentials (de Araujo et al., 2002, Kober et al., 2012, Lee et al., 2007). Rats and humans are able to navigate in visual virtual environments (Höllscher et al., 2005, Gillner et al., 1998). Neural recordings from the hippocampus are possible during functional neural imaging (Dombeck et al., 2010) and even from single neurons intracellularly (Harvey et al., 2009). *Drosophila* flying in a simple virtual environment has helped to elucidate a large range of visual performances and visual learning at multiple levels of analysis since more than 30 years(Wolf et al., 1991, Peng et al., 2007) . However, combining flight behaviour in a virtual environment with neural recordings has turned out to be rather difficult in *Drosophila* leading to some correlations between turning behavior and local field potentials (VanSwinderen et al., 2003). Intra- and extracellular recordings during olfactory learning in restrained honeybees helped to understand multiple facets of sensory encoding and neural correlates of memory formation (Menzel et al., 2006, Denker et al., 2010, Strube-Bloss et al., 2011) but instrumental forms of learning related to visual navigation was not possible so far in honeybees. Here we used an air-supported spherical treadmill that allows the stationary walking honeybee to control the visual environment while long lasting extracellular multi unit recordings were performed from mushroom body extrinsic neurons. We aimed for test conditions in which the animal chose between alternative visual structures and searched for neural correlates of this choice behaviour. The recorded neurons are known to change their properties during olfactory learning sometimes after a consolidation phase of several hours(Okada et al., 2007, Strube-Bloss et al., 2011). These neurons receive input from the intrinsic neurons of the mushroom body. They are sensitive to combinations of multiple sensory modalities including visual stimuli (e.g. the PE1 neuron(Mauelshagen, 1993, Rybak et al., 1998)). The Vummx1, a neuron which is encoding a reward related prediction error (Hammer, 1993) and the PE1 are projecting to the lateral horn (Mauelshagen, 1993). The lateral horn neuropile is innervated by descending sensory-pre-motor connections in the cockroach (Okada et al., 2003). Therefore it seems likely that the PE1 is involved in the initiation of motor output on the basis of reward predictions; in short: decision making. Since the visual system of honeybees (and other insects) differs in many aspects from that of humans, it is a rather difficult task to create an immersive VE for a honeybee. For example bees are able to detect UV light and polarized light (von Frisch, 1914, Lindauer, 1959, Menzel, 1975). Therefore it is important to compare the behavioural and electrophysiological studies of decision making in a virtual environment with comparable measurements in a real setting. To record from freely walking honeybees, a chronically implantable plug for extracellular long term recordings in honeybees was developed (utility patent Nr. 20 2012 002 773.5, de Camp, http://register.dpma.de).

Additionally the plug enables a comparison of electrophysiological studies in freely walking honeybees with well-established paradigms in vertebrates like rats and mice (Hussaini et al., 2011, von Heimendahl, 2012).

Material and Methods

Virtual Environment Setup. Spherical treadmill, geometry of the virtual environment and overall setup were described by Höllscher et al.(2005), apart from the following changes:
The treadmill consisted of a Styrofoam ball (10 cm diameter) placed in a half-spherical plastic cup with several, symmetric located holes through which a laminar air flow passed and let the ball float on air. Laminarity of the air flow was supported by a long (12 m) tube from the well regulated fan to the ball. Low static electrification of the ball and humidity for the animal was achieved by blowing the air into a box with water. Buoys prohibited corrugation inside the box.
The beamer Epson EMP-TW 700 (digital scanning frequency: pixel clock: 13,5-81 MHz, horizontal sweep: 15-60 kHz, vertical sweep: 50-85 Hz, Fourier analysis (15.2.2010): main frequency range 210Hz, less than 10% 100Hz and 60Hz, respectively) was positioned above the Faraday cage and illuminated the inner surface of a cone shaped screen (height 60 cm, bottom diameter 7 cm, top diameter 75 cm) via a large surface mirror and a Perspex window (Fig.1). The inner surface of the cone consisted of white paper. Patterns projected onto this screen were distorted such that they appeared undistorted when seen by the bee (BeeWorld, programmed by Sören Hantke).

During an experiment, the Faraday cage was closed. A webcam (Logitech, Morges Gesellschaft) positioned above imaged the head of the animal via a 500 mm mirror objective allowing observation of the animal during the experiment.

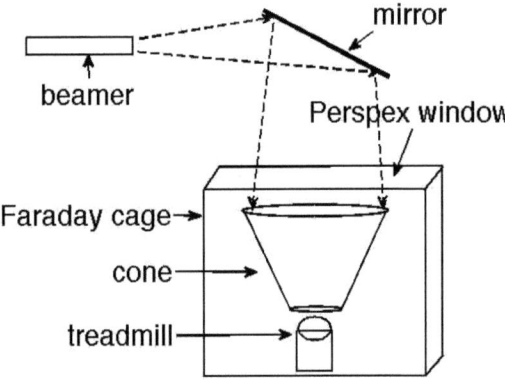

Fig.1: Setup construction. The beamer light is projected onto a cone shaped screen via a mirror through a Perspex window on top of the Faraday cage. The cone can be lifted to place the animal on top of the air supported treadmill.

Control of the virtual environment and experimental procedure. The virtual environment and the recognition of the bees movement was under the control of the customs written program BeeWorld (Sören Hantke). It was implemented in Java by using OpenGL-Bindings for Java (LWJGL). The movement of the ball, initiated by the walking bee, was detected by two optical computer mice as they are used for computer games (Imperator,

Razer Europe GmbH; G500, Logitech Europe S.A.). The mice were accurately positioned under 90° at the equator of the Styrofoam ball and precisely aligned with x/y micro drives. The bee was able to control the virtual scenery by rotatory movements of the ball. Translatory feedback was not used in the experiments described here. Multiple scenarios were programmed. They were realized as xml-files containing the positions, width and color (RGB) of a variable number of vertically oriented stripes. These stripes were positioned around the bee. Scenario7 consisted of alternating black and white, vertically oriented stripes with an angular width of 20° and a grey interspace of 40°. In the figures presented here, the redundant pattern is standardized on one black and one white pattern in an angle of 120° in such a way that 5° in the normalized plot are representing 15° in the virtual environment.

Scenario12 consisted of one black, one white and four blue stripes with the same dimensions as in scenario7.

The field of view of a camera in OpenGL is limited to 179°, the scenarios projected onto the screen, however, simulated a 360° world. In the BeeWorld program a four texture renderer (texturerenderer) was used to create a 360° camera. Data of walking traces were synchronized with the data from spike recordings which were collected with an analog to digital converter (micro3, CED, Cambridge Electronic Design, 30 kHz sampling frequency per channel). A Silicon NPN Phototransistor (BPY 62) directed at the screen detected a short light signal under the control of the BeeWorld program and fed it into the ADC input of the analog to digital converter. The pulse was recorded in a spike two channel and allowed for precise timing between the spike and walking data, recorded with different computers.

A skyline in the background of the scenario consisted of vertically oriented stripes of different hues of grey and different length. A checker board pattern was projected on the plain floor around the bee. The angular rotation of this checker board pattern was equal to the angular movement of the ball simulating a respective movement of the floor directly below and around the bee. The angular rotation of the stripe pattern was set to 75% of the checker board pattern, and a skyline pattern projected onto the screen together with the stripe pattern moved with 50% of the checkerboard pattern velocity. Thus these three patterns simulated depth information by creating a signal of different motion parallax as seen by the stationary walking bee.

Electrophysiology in stationary walking bees. Four insulated copper wires (Elektrisola) with a diameter of 15μm each were coiled and fixed with superglue or heated (210° for 3 minutes have been appropriate to fuse the polyurethane insulation of the wires without producing an electric shortage). These techniques increased the stability of the coiled electrode bundles and facilitated tissue penetration.

The best signal to noise ratio was achieved with Teflon insulated silver ground wires with a diameter of 125 μm (WPI, Berlin, D) implanted into the abdomen or Teflon insulated Platinum/Iridium wires of 25μm diameter (advent, Enysham, Oxford, UK) which were thin enough to be implanted in the brain, compound eye or ocelli. The insulation at the tip of these wires was removed mechanically with a fine forceps.

The tip of the tetrode was cut with a fine scissors. The single ends of the copper wires were dipped into hot solder to remove the insulation and then connected with silver conductive paint (electrolube) to the female side of the pins of an IC socket. After drying, the coiled ends of the electrodes were electroplated using the low-impedance plating procedure with gold and PEG (Ferguson et al., 2009) using the electroplating device NanoZ (Neuralynx). A scanning electron microscope picture of such a partially plated tetrode was kindly provided by Prof. H. Hilger (Fig.2).

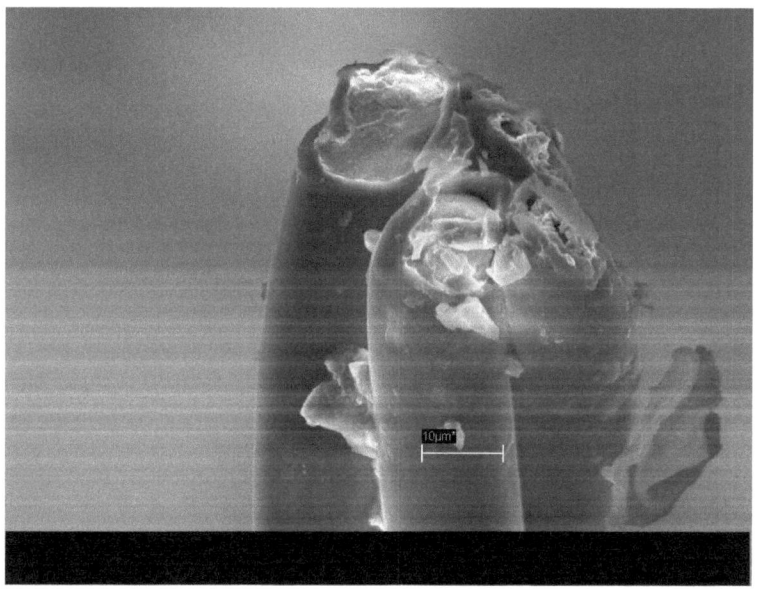

Fig.2: A scanning electron microscope picture from a partially plated tetrode. The polyurethane insulation was melted by heat to fix the coiled structure of the electrode bundle. Both electrodes on the right side show a wobble-like surface texture, due to the gold-PEG plating procedure (Ferguson et al., 2009) . The scale bar indicates 10µm. (Picture kindly provided by H. Hilger)

Preparation of the animals. Bees were caught at the hive entrance or during winter time in a flight cage in a green house. The animals were chilled on ice and fixed temporarily in a small tube with modelling clay. A small piece of plastic tube or rubber foam was fixed with dental wax on the thorax as holder for the stationary running animal on the treadmill. A window between the compound eyes, antennae and ocelli was cut into the head capsule. The tip of the electrode was fixed to a fine forceps which was mounted on an external micromanipulator. The electrode was inserted into the brain, while the animal was still harnessed in the tube. After placing the electrodes in the selected brain area (ventral aspect of the alpha lobe of the mushroom body) under visual control, the electrode was fixed with two component silicone elastomer (kwick sil, WPI) onto the brain and the head capsule. After hardening of the kwick sil, the electrode was released from the external micromanipulator and additionally fixed inside a small slit in the plastic tube/rubber foam holder on the thorax by forming a small loop from the head backwards to the thorax. About 5 minutes later, the bee was released from the tube by grabbing the plastic tube/rubber foam mounted onto the thorax with a forceps and pushing the head with another forceps slowly backwards to facilitate the animals release from the tube. The rubber foam had a slit and the electrodes could be fixed without silicone elastomer. Fixation of the electrodes, however, was crucial for stable and long lasting recordings. Afterwards the bee was clipped by the plastic tube or piece of rubber foam on its thorax to an alligator clip attached to the electrode holder. Longer electrodes were additionally fixed with modelling clay to the electrode holder to prevent reachability of the fine wires from the range of the bee's legs. Especially during the transfer to the setup after the release from the fixation tube and during the first minutes on the ball the animals often tried to remove the electrodes. The electrode holder with the bee was rotated 90° and the bee was carefully adjusted onto the floating ball. The electrode holder consisted of a small balance which kept the animal on the treadmill with its own weight. This balance allowed the animal also to change the distance to the surface of the treadmill during walking. The direct light from the LCD projector was shaded in order to prevent direct illumination of the dorsal regions of the compound eyes and the ocelli. Two UV diodes were positioned within the shade just above the head of the animal simulating short wavelength light coming from above. The micromanipulator and binoculars necessary for positioning the tetrode during the preparation procedure were removed from the setup before the experiment started. Another manipulator allowed precise positioning of the animal on the spherical treadmill.

Extracellular recordings in freely walking honeybees in a Y-maze. The electrodes are constructed as described in the methods for the recordings in the virtual environement. The coiled wires and the ground wire were fixed with silicone elastomer (kwick sil, WPI) to a piece of plastic tube (approximately 10 mm length). The plastic tube was fixed with dental wax onto the animals thorax before. The plastic tube ensured that the recording wires are out of reachability for the bee's legs. The PerspexY- maze had a length of approximately 15cm, the decision point was at about 10 cm from the starting point of the bee. On the floor of the maze a blue and yellow color pattern led to the different arms. The colors were changed during an experiment to differentiate between color learning and mere side preference. In the inter trial intervals the bee was placed on a spherical treadmill without visual feedback.

Extracellular recordings with chronically implanted electrodes. The functional core of the newly developed micro plugs for chronic implantation of extracellular electrodes into the brains of small insects of approximately honeybee size are at least two pieces of plastic tubes (Portex polythene tubing, Smiths Medical, Kent, UK, inner diameter (id) 28μm, outer diameter (od) 61μm, length (depending on the animal) for honeybees about 1.5mm). The polythene tubes inner surface was covered with a conducting fluid by capillary forces (silver conductive paint, Electrolube or Graphit 33, CRC Industries Europe NV). After drying, the quality of the conductive cover on the inner surface was controlled visually.

To construct the smallest plug for a differential extracellular recording three pieces of polythene tubing with a conductive inner surface were needed. Two copper wires (18μm diameter, polyurethane insulation, Elektrisola) were coiled as described above. The coiled wires were cut into pieces of approximately 2 cm length. One end of these small pieces of coiled wires were uncoiled carefully with a fine forceps. The polyurethane insulation of these uncoiled ends was removed by heat (solderer, 400°C). Each single end of the coiled copper wire was inserted in one piece of polythene tubing with inner coating and fixed with a silver glue. For appropriate grounding it was important to have a large metal surface in tissue contact. Therefore a Pt90/Ir10 wire with quadruple Teflon insulation was used (id 25μm, od 35μm, Advent). The insulation was mechanically removed with a forceps at both ends of the PtIr wire. The wire was fixed in a piece of polythene tubing as described for the copper wires. The ends of the tubes with the inserted wires were insulated with superglue. Thereafter the tubes were fixed as a bundle with superglue in such a way, that the open ends of the two recording wires (copper) showed to the same direction and the PtIr wire (ground) open end in the opposite direction. A 2x2x2mm piece of foam rubber was used as bearing for the tubes and wires. The foam rubber was trimmed such that the tubes were sunk in a groove and the other side of the foam had a concave form, oppositely to the convex form of the bees thorax (for a perfect fit when the plug was fixed on top of the bees thorax). From this bee attachment side of the foam, it was penetrated with a fine forceps and the three wires were pulled through the foam. This procedure was necessary to attach the wires near to the bee's body. Otherwise the bees were able to destroy the thin wires with leg movements. The conglomerate of tubes and wires was fixed onto the piece of foam rubber with super glue and the whole block was fixed with dental wax onto the bees thorax. The PtIr ground wire was implanted in one eye or into the abdomen. Careful fixation of the wires onto the head capsule with silicone elastomer (kwick sil, WPI, Sarasota) was important for long lasting recordings without movement artefacts.

The second plug, to connect the chronically implanted electrodes with the preamplifiers, was build by an IC socket with three long copper wires (40cm length, 18µm diameter, polyurethane insulation, Elektrisola). The ends of the long copper wires were dismantled with a solderer (as described previously) and fixed each with silver conductive paint and hard wax to a 1cm long piece of Teflon insulated silver wire (125µm diameter) which was mechanically dismantled at both ends beforehand. The three resulting silver wire ends were fixed to each other with silicone elastomer (kwick sil, WPI). The bare ends were formed like a bind rail with two wires arcuated in one direction and one (the ground connection) to the opposite direction in a way that the three ends geared into each other. In combination with the elastic silicone component this construction guaranteed a stable fitting into the polythene tubes of the chronically implanted electrode device on top of the insect thorax without administering vertical pressure onto the animal. The recordings were performed in a 10x10 cm arena.

Identification of the recording site in the brain. A newly developed extracellular fluorescent copper ion detector Flu TPA1 (Taki et al., 2010) was used to identify the recording site in the brain (Fig. 8c). The dye was solved to saturation in 500 µl PBS buffer (NaCl 137mM, 2.7mM KCl, 8mM Na_2HPO_4, 1.4mM KH_2PO_4, pH: 7.2) for each brain. The freshly removed brain was exposed for 20 to 40 minutes to the dye and directly afterwards fixed in paraformaldehyde (PFA, Electron Microscopy Sciences) for 5-8 hours. Subsequently, the brain was washed for each 10 minutes in PBS, 50% ethanol (EtOH), 70% EtOH, 90% EtOH, 99% EtOH and three times 100% EtOH to remove the water from the brain tissue. To clear the tissue for confocal microscopy it was transferred to a 2:1 mixture of benzyl benzoate and benzyl alcohol. Confocal scans were collected with a Leica TCS. The excitation wavelength of Flu TPA1 was 470 nm, the emission wavelength 510 nm.

Pretraining of freely flying bees. A group of free-flying honeybees, *Apis mellifera L.*, were trained to collect 30% sucrose solution in a feeder located at 10m from the hive. 4 Individually marked foragers were selected from the feeders and trained to enter an experimental tunnel to collect 50% sucrose solution from then on. Only one bee was trained at a time. The experimental setup consisted in a tunnel made out of UV-transparent Plexiglas. The set-up allowed to control the color of the sidewalls and ensured day light condition inside the tunnel. During a pretraining phase bees had to learn to walk to the end of the tunnel to collect a sugar reward. During this phase the floor and sidewalls were covered with unstructured white paper outside the tunnel. Once bees had learned to collect a sugar reward at the end of the tunnel, they were differentially trained to associate a defined arrangement of colors on the sidewalls to the encounter of either positive or negative reinforcement. An experimental bee running towards the reward experiencing color A at the right wall and color B at the left wall was rewarded. If the opposite color/side arrangement was shown . If subjects (color B at the right and A at the left side), the bee was punished with 1 M KCl. After reward or punishment the bee was gently removed from the end of the tunnel and allowed to fly back to the hive. Thus the experimental bee did not experience the opposite color configuration walking towards the entrance of the tunnel. The positive and negative reinforced configuration was presented 10 times each in a pseudo-random manner. After 20 training trials bees were transferred to the virtual environment and their response to the training conditions was tested.

Analysis. Spike2 (Cambridge Electronic Design) Software was used for data sampling and spike sorting. Before sorting the data were high pass filtered. Matlab (2010b, MathWorks) was used for statistical analysis, the calculation of the Fano Factor (Nawrot et al., 2008) and the coefficient of correlation (Spearman's Rho). The mean and variance underlying the Fano Factor were calculated applying a sliding window. The spike times were binned to one second. The variance and the mean of spike rates were calculated for a range of 5 seconds with a sliding factor of 1s. The Fano Factors were calculated for these windows of 5 seconds The distribution of the data was estimated using the Kolmogorov-Smirnov test, the Lillifors test and the Jarque-Bera test. The data presented in this paper are not normal distributed. Therefore non parametric tests were used, like the Wilcoxon ranksum test to compare two variables or the Kruskal-Wallis test to compare more than two groups. The result of the Kruskal-Wallis test, a stats matrix, was further used to calculate pairwise comparisons with Multiple Comparison tests for different confidence levels. The mean plus standard deviations are shown for the spike rate data. Due to the high sample size in these cases, the mean plus standard deviation was a valid method to approximate that 68% of the data are in the interval of mean plus one time standard deviation, 95% of the data are found in the interval of mean plus two times standard deviations and 99% of the data are found in the interval of mean plus three times standard deviation.

Results

Extracellular recordings in the virtual environment. In this study we focused on the analysis of neural recordings and simultaneously recorded walking traces of bees exposed to a 360° panoramic pattern of alternating black and white stripes as described in the Methods section. Three levels of distances were simulated by different angular coupling of the rotatory components of the walking bee, direct velocity coupling of the horizontal checker board pattern in near vicinity to the bee (ground structure), 75% angular velocity coupling of regularly spaced vertical black or white stripes (20°) on light grey background (objects), and 50% velocity coupling (in relation to the animal's walking speed) of the background pattern (skyline) as described in the methods. This VE did not allow for a control of the translatory component by the bees movements. Thus, the animals are able to rotate the scene and select certain areas to the frontal projection field but can not zoom in. Bees exposed to such an environment often showed initially high running activities with little rotatory control of the visual patterns. After several hours the animals started to respond to the objects (Fig. 7b) by temporarily fixing one or the other edge of the objects, and then switching to another edge or object. We shall call this behaviour edge walk (Fig. 6a). Edge walk is characterised by fast turning movements and rest phases before and after selecting the contrast border. These edge walks are not necessarily the dominant walking pattern in a recorded bee over time. Nevertheless, especially in virtual environments with blue stripes the bees were orienting themselves toward stripe edges as shown for pooled duration of stay along four consecutive days in Figures 14a,b and 15a,b.

The recorded neural activity starts more than ten seconds before the turning behaviour and lasts for several seconds after finishing the turning manoeuvre. The spike activity was rather constant and the recording had a good signal to noise ratio during the 4 days of this experiment (Fig. 6c).

In Figure 7a the standardized spike rate during 15 consecutive phases of edge walks is illustrated. An edge run is characterized by a turning (rotation) movement towards a contrast (here the border between a white and a black stripe). The resulting movement pattern with fast rotation towards the contrast area and resting phases until the next rotation movement looks like one or more steps (Fig.6a). The light grey bar in Figure 7a (rotation) is representing the mean rate of 15 edge runs in the same animal during the turning period. The spike rate per turning manoeuvre was divided by the duration of this behaviour. The dark grey bars on the left (Fig. 7a) are one second bins of the spike rate before the turning manoeuvre took place. Every bin represents a mean over 15 consecutive edge walk events.

The same applies for the phase after the turning behaviour, which is represented by the medium grey bars. The horizontal line (Fig. 7a) shows the mean frequency for unit1 during a non edge walking phase. The other dashed horizontal lines represent the mean plus one, two or three standard deviations (SD). The spike rate increases more than 5s before the turning behaviour started. The peak frequency of about 5Hz was reached after the turning behaviour closed, when the animal was resting again. Afterwards the spike frequency decayed slowly (Fig. 7a).

In Figure 7c a statistically significant correlation (Spearman rank correlation, p=1e-13, rho=0.96) between the spike frequency and the amount of rotation of this drone is shown. In contrast to the rank correlation between the spike rate and the rotation, which is more or less constant over days (Figure 3, p=2.6e-6; rho=0.91), the rank correlation between the spike rate and the forward running activity is markedly increased from the second day (Figure 4, p=0.0013 ; rho=0.63) to the fourth day (Figure 5, p=8.7e-13; rho=0.95). Additionally, the rank correlation between the rotatory activity in the virtual environment and the forward running component (translation) is increased from the second day of recording (not shown, p=4.15e-6, rho=0.8) to the fourth day (not shown, p=1.4e-6, rho=0.97).

The bigger amount of rotation and spike rate for the 0° and 120° bins is due to the starting point of the virtual environment at these sites.

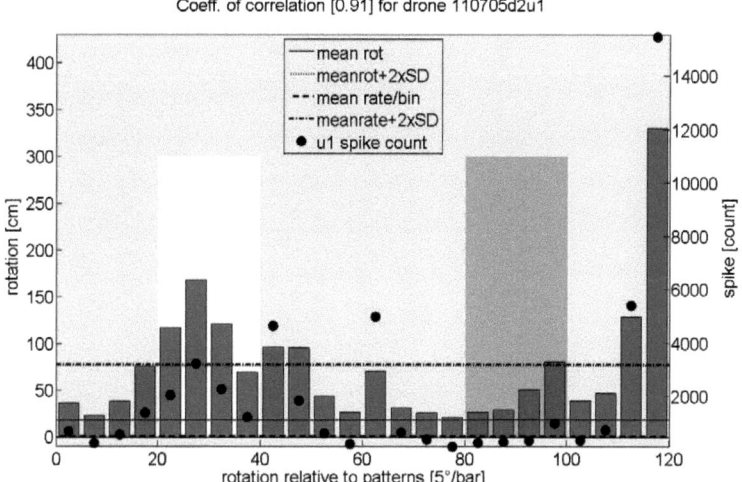

Fig.3: Spearman rank correlation between unit1 rate and rotation. The correlation between rotatory activity in drone 110705 at the second day of the experiment is statistically significant (p=2.6e-6; rho=0.91). The background pattern of the figure is indicating the virtual environment. In this case alternating white and black stripes. The rotation activity was standardized to the smallest part of the redundant pattern (120°). The solid black line indicates the mean rotation (rot), the dashed lines are indicating the mean duration of stay plus two times standard deviation (SD). Please note, that only the left y-axis (rotation) is starting at -10. The right y-axis starts at 0.

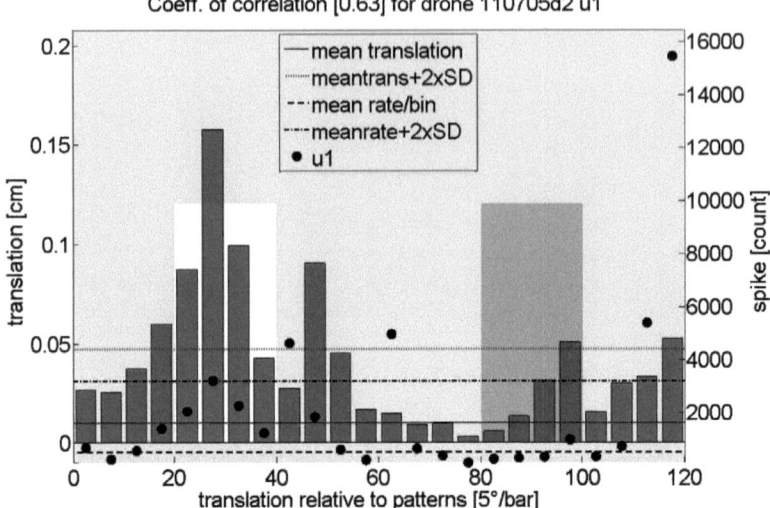

Fig.4: Spearman rank correlation between unit1 rate and translation. The correlation between translatory activity in drone 110705 at the second day of the experiment is statistically significant (p=0.0013; rho=0.63) but not very strong (rho between 0 and 1, with one for total correlation). Note the higher amount of forward running on the white stripe. The dark grey bars are representing the translation for 5°, respectively (but 15° for the 360° environment, due to standardization on 120°). The background pattern of the figure is indicating the virtual environment. In this case alternating white and black stripes. The translation activity was standardized to the smallest part of the redundant pattern (120°). The solid black line indicates the mean translation (meantrans), the dashed lines are indicating the mean duration of stay plus two times standard deviation (SD). Please note, that only the left y-axis (translation) is starting in the negative range. A negative translation would indicate backward running.

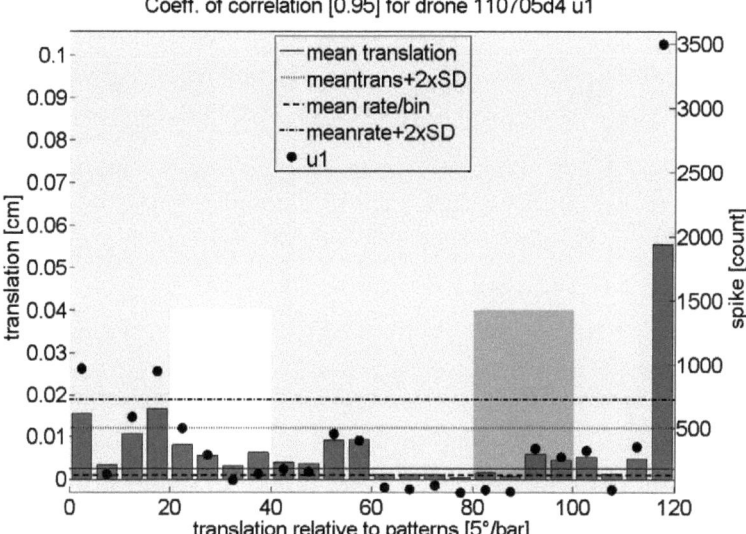

Fig.5: Spearman rank correlation between unit1 rate and the translation. The correlation between translatory activity in drone 110705 at the fourth day of the experiment is statistically significant (p=8.7e-13; rho=0.95). Note the decrease of forward running in comparison to the second day. The dark grey bars are representing the translation for a range of 5°, respectively (but 15° for the 360° environment, due to standardization on 120°). The background pattern of the figure is indicating the virtual environment. In this case alternating white and black stripes. The translation activity was standardized to the smallest part of the redundant pattern (120°). The solid black line indicates the mean translation (meantrans), the dashed lines are indicating the mean duration of stay plus two times standard deviation (SD). Please note, that only the left y-axis (translation) is starting in the negative range. A negative translation would indicate backward running.

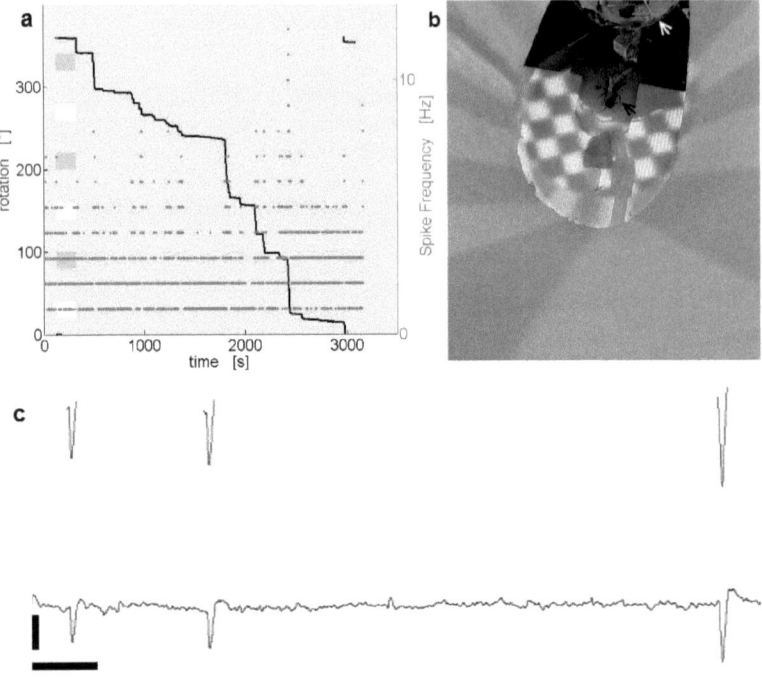

Fig.6: Extracellular recordings in honeybees during virtual navigation. **(a)** A typical step run walking pattern. At the left border of the figure the virtual cues are indicated as white and black (plotted in light grey) stripes. The rotation (black trace) is plotted on the left y-axis. The spike frequency of unit1 (grey dots) is plotted on the right x-axis. The step like pattern is due to fast rotations and pauses in between. **(b)** A top view into the virtual environment during an experiment. The black arrow is pointing toward the running bee and the white arrow toward the light shade. **(c)** Signal to noise ratio of the recording. The bottom trace is the original recording, an example of the sorting result is shown on top. The vertical scale bar indicates 500µm, the horizontal bar a time range of 0.01s.

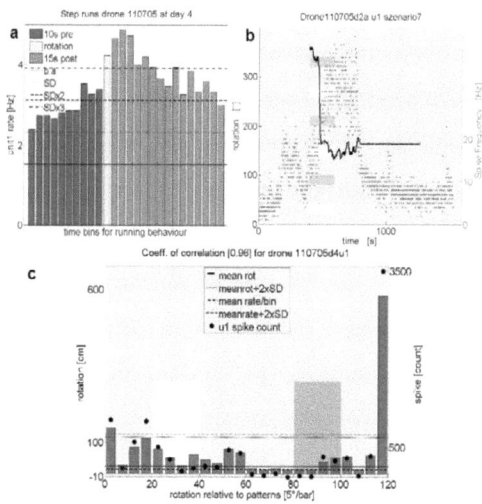

Fig.7: Neural correlates of virtual navigation. **(a)** Spike activity for unit1 (u1) during 15 consecutive "step runs". Each count of spike activity during the turning movement is standardized to its duration (light grey bar). 10 seconds before and 15 seconds after the spontaneous turning took place, the spike rate is pooled in one second bins (dark grey bars before (pre) and middle grey bars after (post) the light grey bar in the middle). The solid line indicates the mean spike rate during non step walk navigation (baseline activity – b.a.). The dashed lines indicate the mean spike activity plus one, two and three times standard deviation (SD). The spike activity was increased 10s before the turning behaviour took place and reached a peak activity shortly after the turning movement. ...

...Fig.7: **(b)** Neural activity of unit1 and rotatory activity (rot) in the same drone. The start of the virtual scene is indicated by the grey and white bars (grey bars have been black during the experiment). Please note the correlation of active navigation, target selection and spike activity (dots, right y-axis). **(c)** Statistically significant Spearman rank correlation between rotation and spike activity for unit1 at day four (rho= 0.96, p=1e-13). The lines indicate mean and mean plus two times standard deviation (SD) for the spike activity and the rotation. The right and left borders of the figure are the starting points of the virtual scene. The data are standardized to 120° of the 360° panoramic virtual environment.

Recordings in pretrained bees in the virtual environment. In a second approach in the virtual environment it was tested whether learned visual patterns during free flight are also recognized during virtual navigation. Freely flying honeybees were trained (thanks to Jaime MartinezHarms) to enter a Perspex tunnel if the CS+ (conditioned stimulus) color combination was displayed on the sidewalls inside the tunnel (CS+: blue left, yellow right; CS-: yellow left, blue right). At the end of the tunnel, the bees were rewarded with sucrose solution, in case of the CS+ color combination. Bees did not get any reward if they entered the tunnel in the CS- color combination (for a detailed description see methods). After one day of free flight training with several trials, the marked bees were caught at the end of the tunnel and transferred to the virtual environment. Preparation and electrode implantation were done as described in the methods. The CS- and CS+ patterns were presented in a pseudo-randomized order in the virtual environment. CS+ trials were rewarded after recording the running velocity for a time range of 30-60s to prevent extinction effects (Bittermann et al., 1983).

For analysis, we split the bees in two groups. The learner bees (n=2) learned the discrimination task during free flight and virtual navigation. These bees showed a statistically significant faster forward running velocity to the CS+ than to the CS- color combination in the virtual environment (Wilcoxon ranksum test:p=0.0315, n=6 trials) (Fig. 3a). The Non learner bees (n=2) belong to the second

group. They did not tranfer the visual discrimination task to the virtual situation. There is no statistically significant difference between the running velocity of the non learner bees towards the CS+ or the CS- color condition (Wilcoxon ranksum test: p=0.8518, n=6 trials) (Fig. 8b).

For one learner bee (110904J22) the bursting intensity is significantly different in unit1 and 2 for the rewarded and unrewarded stimulus situation in the virtual environment. These two units are spiking antagonistically in bursts. For unit 1 the Fano Factor is statistically significant reduced in the CS- (Multiple comparison test based on Kruskal-Wallis test results (p=4.5e-23): Confidence Interval (CI) for 99% Confidence Level [140.8 : 264.5]) and the CS+ (Confidence Interval (CI) for 99% Confidence Level [42.8 : 195]) color condition in comparison to a control situation in which the bee was recorded inside the virtual arena but without any pattern (Fig.8d). For unit 2 in the same animal during the same trials, the Fano Factor is only reduced in the CS- condition (Figure 9). The Fano Factor was calculated with a sliding window. The raw spike times have been binned for 1second.

The recording site was identified with a fluorescent probe for copper ion detection (Taki et al., 2010) (confocal scan, Fig. 8c). The scale bar is indicating a range of 100μm. A black arrow is showing the site of the electrode tip at the ventral border of the alpha lobe, a part of the mushroom body.

The white area is the stained brain tissue, where copper ions could be detected.

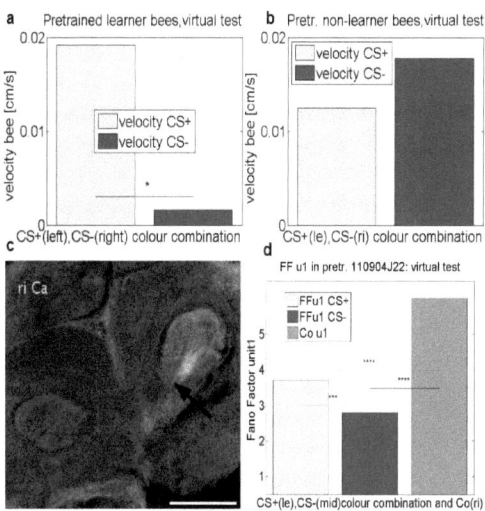

Fig.8: Shift from real world memory to virtual navigation. During free flight the bees learned to enter a tunnel with a specific colour pattern on the side walls (CS+). The CS+ colour combination was rewarded with sugar water. The CS- situation was negatively reinforced. **(a)** In the virtual test, two of four pretrained bees showed significantly higher forward running velocities if the CS+ colour pattern was projected in the virtual scene (learner bees, Wilcoxon ranksum test, alpha=0.05, p=0.032) against the alternative CS- pattern. **(b)** In the non learner bees no such effect existed. **(c)** The site of recording in the brain tissue was stained via a fluorescent copper ion detection dye, suitable for confocal microscopy (Taki et al., 2010). The black arrow points toward the recording site at the ventral border of the left alpha lobe, a part of the mushroom body. The right calyx is labeled for orientation (ri Ca). ...

...Fig.8: **(d)** In one learner bee, the burst activity (measured as Fano Factor (FFu1)) of unit1 was significantly reduced for the CS+ (Multiple Comparison test, based on Kruskal Wallis result (prob.>Chi-sq.=4.5e-23), confidence level (CL)=99%, confidence interval (CI)=[42.8 - 195]) and the CS- (Multiple Comparison test, CL=99%, CI=[140.8 – 264.5]) trials in comparison to a control (Co) situation without a projected environment. The spike variability was significantly lower in the CS- situation in comparison to the CS+ trials (Multiple Comparison test, CL=97.5%, CI=[1.4 – 166]), (le=left, ri=right, mid=middle).

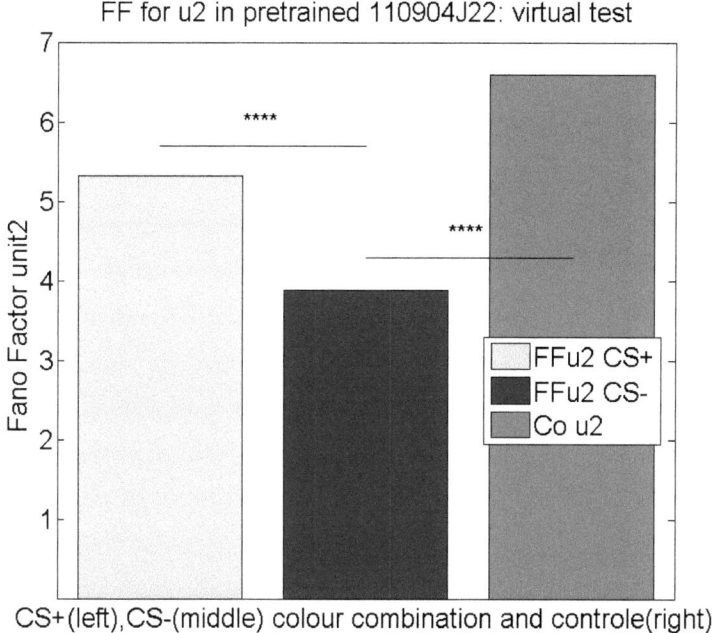

Fig.9: Bursting activity of unit2 during the virtual test in a pretrained honeybee. The Fano Factor (FF) is shown as measure of spike variance to show bursting activity of unit2 (u2) spikes during seven (4xCS+, 3xCS-) consecutive tests of one pretrained honeybee in the virtual environment. The bursting activity was statistically significant reduced in the CS- test in comparison to the control situation without any pattern but inside the virtual arena (Multiple comparison test based on Kruskal-Wallis test results (p=9.4e-9): Confidence Interval (CI) for 99% Confidence Level (CL) [67.2 : 191.2]). ...

...Fig.9: ...The same applies for the comparison of CS+ and CS- situation. The bursting activity of the CS- situation was significantly reduced in comparison to the CS+ situation (Multiple comparison test based on Kruskal-Wallis test results (p=9.4e-9): (CI) for 99% CL [11.9 : 195.9]).The bursting activity during the CS+ situation was not statistically significant different from the control situation in the virtual environment (Multiple comparison test based on Kruskal-Wallis test results (p=9.4e-9): CI for 90% CL [-28.2 : 78.7]). (n for FF in different groups: Control: 486, CS+:53, CS-:84, sliding window for calculation of FF, see methods).

Extracellular recordings during decision making in a Y-maze. The spike activity of two recorded units during 18 consecutive spontaneous visual decisions in a Perspex Y-maze with coloured floor patterns was pooled. The bees showed a side preference rather than a color learning.

The frequency for unit1 and unit2 was elevated during the decision period in comparison to a non-decisive control situation with stationary walking (Fig. 10). Unit1 activity has increased above the mean plus three times standard deviation level of a control situation during the whole calculated time period of the decision process from one second before the decision until three seconds after the decision.

Unit2 showed increased (above the level of mean control spike activity plus three times standard deviation) spike frequencies for the first two seconds after the decision for one of the arms of the Y-maze.

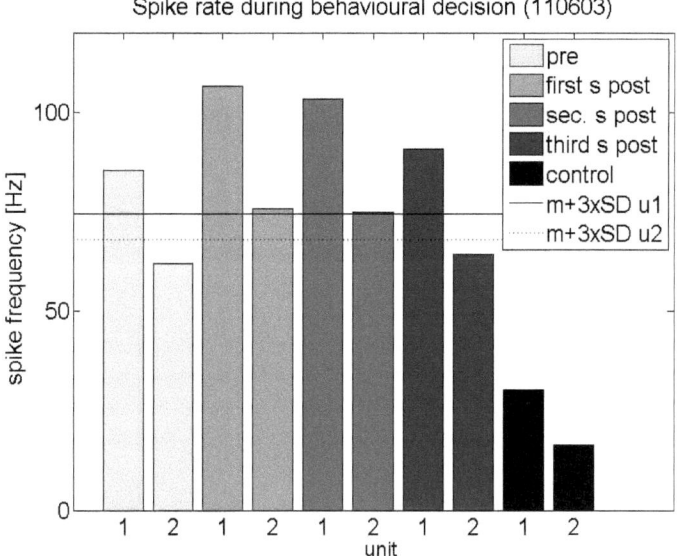

Fig.10: Neural correlates of decision making. One freely running bee was recorded during 18 consecutive decisions in a y-maze. The bee did not learn a color pattern presented on the ground but developed a side preference. Two units were recorded. Unit1 (u1) showed an increase in spiking activity from 1s before the decision (pre) till three seconds after the decision (post). Unit2 (u2) revealed increased spiking activity in the two seconds after the decision. In the control situation the bee was walking stationary on a ball, without a decision task. The mean plus 3 times standard deviation (SD) thresholds for unit1 and 2 are indicated as solid and pointed line, respectively.

Extracellular recordings with chronically implanted micro plugs in freely running honeybees.

The brain potentials were recorded for two consecutive days and one night. The bee lived with the plug for six days. As proof of recording quality, an optical neuron was recorded which spiked with the 50 Hz flicker frequency of the room light (Fig. 11d). The differential recording electrodes were placed in the mushroom body. One Unit disappeared during the first day. A second unit was stable for both days and a third unit was present for short interspersed time ranges in both days.

During the first day of recording defined regions of elevated spike frequency (bursts) for unit1 and unit2 are seen. These periods were characterised by a short baseline shift at the beginning and a negative baseline shift at the end of the high frequency period (Figure 12). The periods had random duration (up to a few seconds) and random inter period intervals. The high frequency periods occurred the first time when the animal started to explore the arena after recovery from the preparation procedure. During the first night, the animal showed prolonged phases of inactivity (here defined as sleep). The high frequency periods could have been observed during sleep and during active exploration of the arena. 15 minutes after the last sleep phase, the high frequency periods disappeared. Figure 13a shows a PSTH for five such high frequency periods during sleep at night. The bursts starting points were aligned at second zero. The last high frequency bursts after the sleep period are shown in Figure 13b.

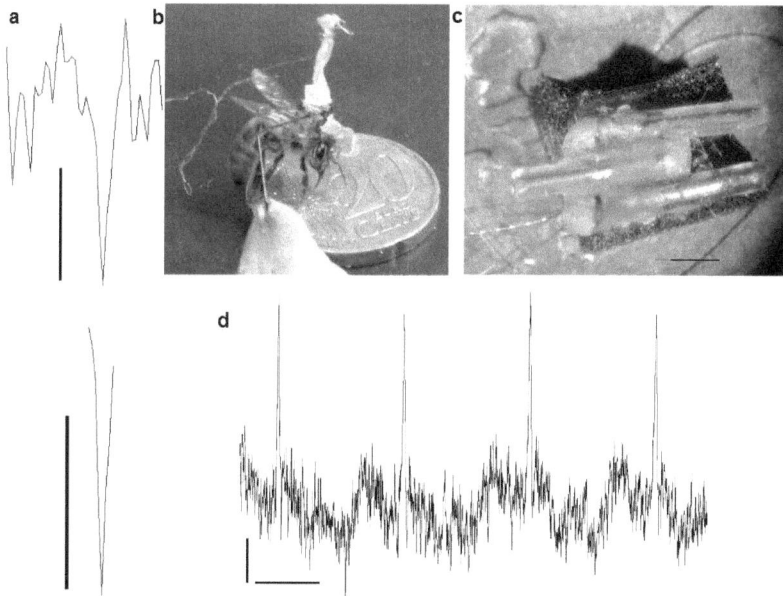

Fig.11: Extracellular recordings with a chronically implanted micro plug. (a) An original spike, recorded in the mushroom body of a navigating honeybee (vertical scale bars=10μV) and the spike extracted by a spike sorting algorithm (below). The horizontal scale bar indicates a time range of 0.001s. (b) Honeybee with a chronically implanted electrode on top of a 20 Euro Cent coin. The long component on the thorax is the removable connecting device from the chronically implanted electrodes to the preamplifiers. (c) The chronically implanted part of the plug system on top of a 20 Euro Cent coin (the scale bar indicates 1mm). The black part was fixed on top of the honeybees thorax. (d) Spikes, recorded with the micro plug system. This optical neuron was firing with the 50Hz room light flicker (horizontal scale bar=0.01s, vertical scale bar=50μV).

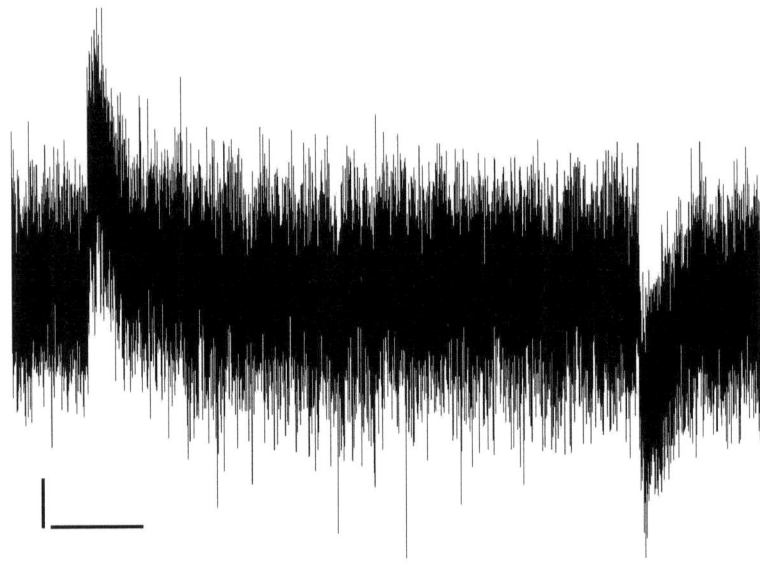

Fig.12: Ripple-like events during exploratory behaviour and sleep. The periods of high spike frequency were characterized by an initial positive baseline shift and a final negative baseline shift. The ripple-like shift phases had a frequency of approximately 10 Hz. These phases were visible at the first day when the animal recovered from the preparation procedure and started with exploration of the 10x10 cm arena. The events were seen during sleep in the following night but lasted only for additional 15 minutes after the last sleep phase (sleep was characterized by the absence of movement for up to several minutes). The vertical bar indicates 4 µV, the horizontal scale bar indicates 0.2s. The data have been recorded with the chronically implanted micro-plug.

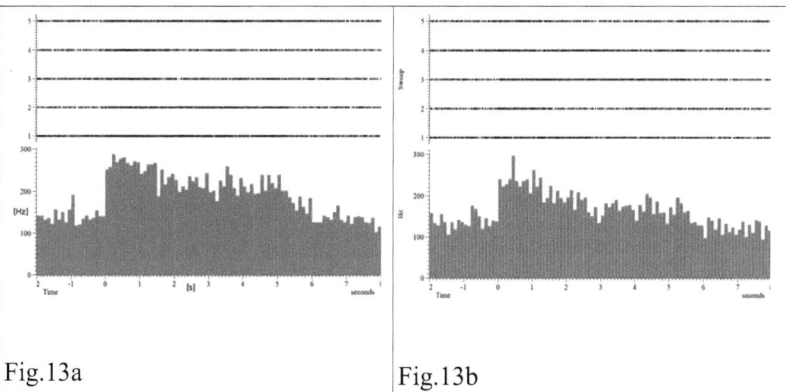

Fig.13a Fig.13b

Fig.13: a)Bursts in the mushroom body of a freely navigating honeybee during sleep (night recording) after exploration. A bee with a chronically implanted micro plug was recorded during exploration in a small arena. During early exploration the animal developed burst states which were characterized by special shifts in the baseline but varied in duration and inter burst interval. These reverberating bursts were also recorded during a sleep phase at night, as shown in this figure. The unit 2 bursts are aligned with their starting points at second zero. Above, the spike events of the single burst events are displayed, which are pooled in the PSTH at the bottom. Note the increase in spike frequency from approximately 150 Hz before the burst onset to nearly 300 Hz peak amplitude at the beginning of the bursts. b)Last occurence of high frequency periods during free navigation after sleep. Five consecutive burst periods are pooled and the starting points aligned at second zero (xaxis). The spike frequency was increased from approximately 150 Hz baseline activity to 250 Hz during the bursts (0.1s bins). Single traces with spike events are plotted in 1a, above the PSTH.

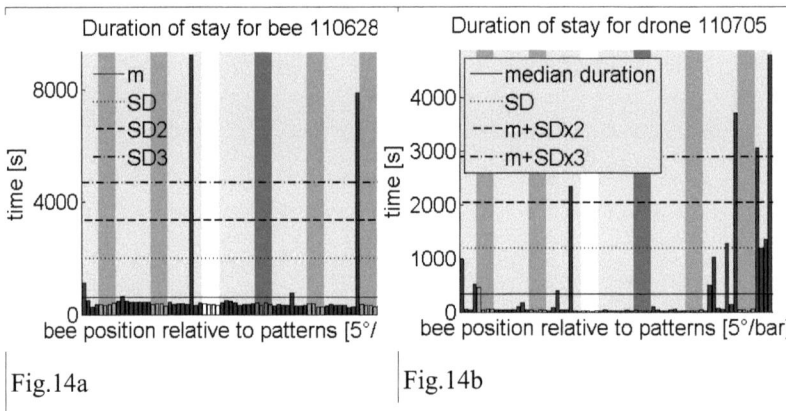

Fig.14a Fig.14b

Fig.14: Duration of stay for a forager honeybee and a drone in szenario12. The light grey bars (indicated in the background of the figure) were blue in the virtual szenario12 and the dark grey bar was black. The place preferences are pooled data for four consecutive days for each bee. The forager bee (110628, a) preferred one locality at the border of a blue stripe (right) and one place in near vicinity to the white stripe. The drone (110705, b) preferred the same places. Additionally the drone had longer durations of stay at the starting point in the virtual environment (the right and left borders of each plot). The solid black line indicates the mean duration of stay, the dashed lines are indicating the mean duration of stay plus one, two or three times standard deviation (SD).

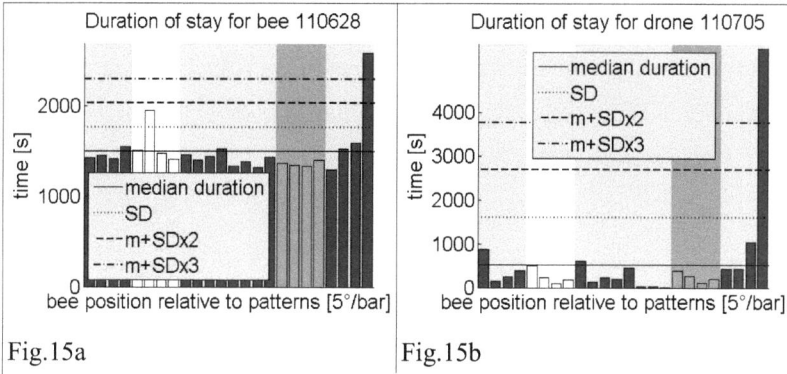

Fig.15a Fig.15b

Fig.15: Duration of stay for a forager honeybee and a drone in szenario7. The grey bar in the background of the figures indicates a black stripe in the virtual szenario7. The pattern consisted of alternating black and white stripes in the virtual environment. One redundant part of this pattern is shown here and the data are standardized to 120°, which mean that 5° are representing 15° of the original szene. Both bees had a clear peak duration at the release site (right border of the figures) and a tendency to prefer the white stripe. The general angular distribution is much more homogeneous for szenario7 in comparison to szenario12. The solid black line indicates the mean duration of stay, the dashed lines are indicating the mean duration of stay plus one, two or three times standard deviation (SD).

Discussion

Spontaneous focusing onto stripe patterns as shown in Figure 7b and associated changes in neural activity in the extracellular recording of the drone indicate that bees are able to navigate in a virtual environment. It was not possible to train bees to select particular visual patterns in the virtual environment. Since 50% of the pretrained bees are able to transfer a learned colour pattern during free flight to a virtually presented colour pattern (Fig. 8a, b), the reason for the absence of learning in the virtual environment can not simply be due to primary sensory conflicts like insufficient visibility of beamer projections for the bee compound eyes. As Karl von Frisch stated (1914), the ability of learning is related to the animals necessities. During free flight training of colour patterns, the bees are allowed to keep their own pattern of task timing. Usually the bees are visiting an artificial feeding station with some kind of cue and suck sugarwater. Afterwards they are flying back to the hive, their social context. This procedure and the ongoing social contact might be important for visual learning or at least the motivational basis for doing so.

Karl Weiss (1954) showed that walking bees and wasps are indeed able to learn rules, associated with colors during navigation in a flat maze. Additionally, different motivational states of the animal may influence the walking pattern in the virtual environment. Gain state influenced changes in optomotor head movements were described for blowflies (Rosner et al., 2009).

„Motivation" seems to be an important point for performing operant conditioning tasks in the virtual environment. An animal which is stressed by the experimental procedure or the preparation might be in the flight mode rather than any learning related behavioral mode. The next step will be to improve the immersive properties of the virtual environment. Especially translatory feedback for the movements of the bee on the floating ball in order to optimise the virtual environment as illusion of a natural situation for the bee. This environment might also include different odours. It has to be elucidated in how far olfactory and visual cues interfere in walking honeybees. For the fruitfly *Drosophila melanogaster* it was shown that olfactory modification of visual reflexes takes place in the mushroom body (Chow et al., 2011). In restrained honeybees odour information seems to be dominant (Niggebrügge et al., 2009). Chemosensory dominance has also been observed for freely flying bees (von Frisch, 1914).

Additional evidence for active virtual navigation of honeybees is seen in the comparison of duration of stay for four consecutive days in two different virtual environments in a forager honeybee and a drone. Both had nearly the same spatial distribution of peak durations of stay for the scene with blue stripes and nearly no peaks in the distribution of duration of stay for the pattern with black and white stripes (Figures 14a,b and 15a,b). Such observations and exactly the imperfect immersive properties of the virtual environment or at least different scenes might help us to learn more about the way bees are recognizing their environment and which conditions are necessary for learning and navigation tasks.

An operant conditioning paradigm for honeybees in the virtual environment is a goal for the future and might be a great advantage in order to study the neural basis of behavioural decision making in an insect model.

The recorded unit1 in the drone presented in Figure 6 and Figure 7 seems to be related to arousal in the broader sense. This unit is active in the pre decision phase (maybe visual orientation by the animal or arousal) and even after the movement has been finished (Fig. 7a). The long pre and post rotation activities of unit1 excludes mere motor related activity. Possibly such activity patterns are used as gateway to recruit other parts of the network in a behavioural defined time range.

Another interesting neural correlate of learning is observed in one pre-trained forager bee. The neural activity did not alter in the total mean frequency but in bursting activity in shorter time windows. During a control measurement, the variance of spike times (which is high during ongoing short burst bouts) was high in comparison to the learned CS+ and CS- trials. Interestingly, the burst activity was lowest for the CS- colour condition (Fig. 8d). These data give evidence, that at least in freely running honeybees, bursting or variabilities in spike variance seems to have coding properties in the field of operant learning or memory. Variability has been described for cortical spike trains (Nawrot et al., 2008).

The virtual environment is a tool to shift the bee navigation to the lab. The second approach is to shift the electrophysiology to navigation in a real environment. Therefore, a micro plug for chronical implantation of extracellular electrodes in the bee brain was developed. Not only that the bees are living up to one week with these electrodes, the plugs are an opportunity to keep the bees own kind of learning rhythm without social deprivation in the conventional lab situation. The combination of both approaches has the great potential to give first insights into the neural basis of behavioural decision making and navigation in eusocial insects. One interesting question for the future would be if neural correlates of place learning are also existing in insects? Figure 13a,b indicates that there are indeed correlates of place learning and consolidation during sleep in insects. Interestingly the high frequency bursts which were observed during exploration behaviour of the bee at the first day are disappearing 15 minutes after the first sleep bout in this environment. Additionally the sharp and characteristic baseline shifts (Figure 12) at the beginning and at the end of these high-frequency spike phases have some similarities with ripples, found in navigating rodents (Buzsáki, 1989). The ripple-like events were in the 10 Hz range. It remains speculative if this is a direct observation of a consolidation phenomenon. Replay events during navigation pauses and sleep are often discussed in the context of consolidation of place meory for vertebrates and associated with ripples (Davidson et al., 2009). In Figure 10 we show neural correlates of decision making in a honeybee, freely running in a Y-maze. The spike activity of unit one is already increased about one second before the decision (to enter one of two arms of the maze), took place. The second unit had higher activity levels after the decision point. It seems that a decision process is temporarily resolved in slightly deferred activity patterns of different neurons. Cells participating in different sequences of a decision trial at distinct points in time have been shown in a recent study in virtually navigating mice (Harvey et al., 2012).

Another interesting point would be the measurement of neural correlates of diseases in individual bees. A recent study raised the question whether the colony collapse disorder in honeybees is due to infections with parasitic arthropods which alter the bees ability to navigate (Core et al., 2012).

Therefore the next step might be a wireless recording with an extended form of the micro-plugs described here.

Why is it important and necessary to study the neural correlates of decision making in insects? Humans and insects decide in similar ways (Louâpre et al, 2010). Since artificial intelligence and biologically inspired robots are a growing field of interest, conceptual insights to brain processes that underlie decision making and navigation might be of interest with respect to the development of biologically inspired algorithms and autonomously acting agents (Lauer et al., 2011, Behnke&Rojas, 2001, Riedmiller, 2005a).

In this paper we presented two combined or individually usable methods for studying the neural correlates of behavioural decision making and navigation in insects of about honeybee size. The data based on the newly established methods described here, give the first evidence for navigational replay events, consolidation of these reverberating neural patterns during sleep and distinct timing of neural activity while performing navigational decisions in honeybees.

References

de Araujo, D.B., Baffa, O., Wakai, R.T. Theta Oscillations and Human Navigation: A Magnetoencephalography Study. J. Cogn. Neurosci. **14**, 70-78 (2002).

Behnke, S., Rojas, R. A Hierarchy of Reactive Behaviors Handles Complexity. In Proc. ECAI2000 Workshop Balancing Reactivity Social Deliberation in Multi-Agent-Systems, 125-136 (2001).

Bittermann, M.E., Menzel, R., Fietz, A., Schäfer, S. Classical Conditioning of Proboscis Extension in Honeybees (*Apis mellifera*). J. Comp. Psychol. **97**, 107-119 (1983).

Buzsáki, G. TWO-STAGE MODEL OF MEMORY TRACE FORMATION: A ROLE FOR "NOISY" BRAIN STATES. Neurosci. **31**, 551-570 (1989).

Chow, D.M., Theobald, J.C, Frye, M.A. An Olfactory Circuit Increases the Fidelity of Visual Behavior. J. Neurosci. **31**, 15035-15047 (2011).

Core, A. et al. A New Threat to Honey Bees, the Parasitic Phorid Fly *Apocephalus borealis*. PloS One **7**, e29639 (2012).

Davidson, T.J., Kloosterman, F., Wilson, M.A. Hippocampal Replay of Extended Experience. Neuron **63**, 497-507 (2009).

Denker, M., Finke, R., Schaupp, F., Grün, S., Menzel, R. Neural correlates of odor learning in the honeybee antennal lobe. Eur. J. Neurosci. **31**, 119-33 (2010).

Dombeck, D.A., Harvey, C.D., Tian, L., Looger, L.L., Tank, D.W. Functional imaging of hippocampal place cells at cellular resolution during virtual navigation. Nat. Neurosci. **13**, 1433-1440 (2010).

Ferguson, J.E., Boldt, C., Redish, A.D. Creating low-impedance tetrodes by electroplating with additives. Sens. Actuators. A Phys. **156**, 388-393 (2009).

von Frisch, K. Der Farbensinn und Formensinn der Bienen. Zool. Jb., Abt. Allg. Zool. u. Physiol. **35**, 1-238 (1914).

Gillner, S., Mallot, H.A. Navigation and Acquisition of Spatial Knowledge in a Virtual Maze. J. Cogn. Neurosci. **10**, 445-463 (1998).

Harvey, C.D., Coen, P., Tank, D.W. Choice-specific sequences in parietal cortex during a virtual-navigation decision task. Nature **484**, 62-68 (2012).

Harvey, C.D., Collman, F., Dombeck, D.A., Tank, D.W. Intracellular dynamics of hippocampal place cells during virtual navigation. Nature **461**, 941-946 (2009).

von Heimendahl, M., Rao, R.P., Brecht, M. Weak and Nondiscriminative Responses to Conspecifics in the Rat Hippocampus. J. Neurosci. **32**, 2129-2141 (2012).

Höllscher, C., Schnee, A., Dahmen, H., Setia, L., Mallot, H.A. Rats are able to navigate in virtual environments. J. Exp. Biol. **208**, 561-569 (2005).

Hussaini, S.A., Kempadoo, K.A., Thuault, S.J., Siegelbaum, S.A., Kandel, E.R. Increased Size and Stability of CA1 and CA3 Place Fields in HCN1 Knockout Mice. Neuron **72**, 643-653 (2011).

Kober, S.E., Kurzmann, J., Neuper, C. Cortical correlate of spatial presence in 2D and 3D interactive virtual reality: An EEG study. Int. J. Psychophysiol. **83**, 365-374 (2012).

Lauer, M., Schönbein, M., Lange, S., Welker, S. 3D-objecttracking with a mixed omnidirectional stereo camera system. Mechatronics **21**, 390-398 (2011).

Lee, J.H., Kwon, H., Choi, J., Yang, B.H. Cue-Exposure Therapy to Decrease Alcohol Craving in Virtual Environment. Cyberpsychol. Behav. **10**, 617-623 (2007).

Lindauer, M. Angeborene und Erlernte Komponenten in der Sonnenorientierung der Bienen. Z. Vergl. Physiol. **42**, 43-62 (1959).

Louâpre, P., van Alphen, J.J.M., Pierre, J.S. Humans and Insects Decide in Similar Ways. PloS One **5**, e14251 (2010).

Mauelshagen, J. Neural Correlates of Olfactory Learning Paradigms in an Identified Neuron in the Honeybee Brain. J. Neurophysiol. **69**, 609-625 (1993).

Menzel, R., Giurfa, M. Dimensions of Cognition in an Insect, the Honeybee. Behav. Cogn. Neurosci. **5**, 24-40 (2006).

Menzel, R. Electrophysiological Evidence for Different Colour Receptors in One Ommatidium of the Bee Eye. Z. Naturforsch. **30 c**, 692-694 (1975).

Nawrot, M.P. et al. Measurement of variability dynamics in cortical spike trains. J. Neurosci. Meth. **169**, 374-390 (2008).

Niggebrügge, C., Leboulle, G., Menzel, R., Komischke, B., Hempel de Ibarra, N. Fast learning but coarse discrimination of colours in restrained honeybees. J. Exp. Biol. **212**, 1344-1350 (2009).

Okada, R., Rybak, J., Manz, G., Menzel, R. Learning-Related Plasticity in PE1 and Other Mushroom Body-Extrinsic Neurons in the Honeybee Brain. J. Neurosci. **27**, 11736-11747 (2007).

Peng, Y., Xi, W., Zhang, W., Zhang, K., Guo, A. Experience Improves Feature Extraction in *Drosophila*. J. Neurosci. **27**, 5139-5145 (2007).

Riedmiller, M. Neural Fitted Q Iteration – first experiences with a data efficient neural reinforcement learning method. In Proc. of the European Conference on Machine Learning, ECML 2005, Porto, Portugal, (October 2005a).

Rosner, R., Egelhaaf, M., Grewe, J., Warzecha, A.K. Variability of blowfly head optomotor responses. J. Exp. Biol. **212**, 1170-1184 (2009).

Rybak, J., Menzel, R. Integrative Properties of the Pe1-Neuron, a Unique Mushroom Body Output Neuron. Learn. Mem. **5**, 133-145 (1998).

Strube-Bloss, M.F., Nawrot, M.P., Menzel, R. Mushroom Body Output Neurons Encode Odor-Reward Associations. J. Neurosci. **31**, 3129-3140 (2011).

Taki, M., Iyoshi, S. Ojida, A., Hamachi, I., Yamamoto, Y. Development of Highly Sensitive Fluorescent Probes for Detection of Intracellular Copper(I) in Living Systems. J. Am. Chem. Soc. **132**, 5938-5939 (2010).

Van Swinderen, B., Greenspan, R.J. Salience modulates 20-30 Hz brain activity in *Drosophila*. Nat. Neurosci. **6**, 579-586 (2003).

Weiss, K. Der Lernvorgang Bei Einfachen Labyrinthdressuren von Bienen und Wespen. Z. Vergl. Physiol. **36**, 9-20 (1954).

Wolf, R.,Heisenberg, M. Basic organization of operant behavior as revealed in *Drosophila* flight orientation. J. Comp. Physiol. A **169**, 699-705 (1991).

5. Are ripples/sharp waves older than the hippocampus?

Extracellular recordings in the mushroom body of eusocial insects give strong evidence for the occurence of sharp waves and place related spiking activity. The occurence of sharp waves/ ripples is associated with replay and learning during slow wave sleep (SWS) or awake pauses during navigation in rodents. Furthermore it was shown that memory consolidation is affected by ripple destruction. Two newly established methods allow for extracellular long term brain recordings in comparably small protostomian animals: insects. Surprisingly the comparison between the mammalian hippocampus and the insect mushroom body seems to be valid on the level of electrophysiology. A freely navigating honeybee and a virtually navigating hornet show ripple-like events with highly synchronized spikes. The ripples are disappearing after sleep during night in the freely navigating bee. These new insights to complex phenomena in the insect brain are not only interesting for comparative brain studies but also for the question when and how such brain potentials evolved. Ripples, replay and associated sleep states seem to be evolutionary far older than the mammalian brain.

Introduction. The firing pattern of place cells in the rodent hippocampus encodes a spatial map of the explored environment (O'Keefe and Dostrowski, 1971; Wilson and McNaughton, 1993). During immobile activity and SWS the neural network of the spatial map is reactivated, embedded in characteristic sharp wave/ ripple potentials in the hippocampus (Buzsáki, et al., 1992). Physiological (Mizunami, 1998) and molecular (Kandel and Abel, 1995) similarities between the Hippocampus and the mushroom bodies have been discussed in the past. Tomer et al. (2010) developed a new technique which allowed for a simultaneous gene expression profiling of the whole brain. This large scale molecular fingerprinting makes comparative studies possible. They tested the homology hypothesis for the annelid mushroom body and the vertebrate pallium. Anatomical studies by Strausfeld et al. (1998) could not find a solution for this hypothesis and were also not sufficient to clarify the homology hypothesis for the mushroom body inside the Tetraconata (monophyletic group of Hexapoda and Crustacea (Friedrich&Tautz, 1995, Regier et al., 2005)). Tomer et al. (2010) found that the vertebrate pallium and the annelid mushroom body develop from the same, molecularly characterized subregion (emx+/pax6+) with resulting similar patterning mechanisms. The statistical significance of homologue pallium and mushroom body development was further supported by unique sets of transcription factors, intrinsic glutamatergic neurons and brain specific microRNAs. Furthermore Tomer et al. found strong evidence for an evolutionary relation between the annelid and insect mushroom body anlagen via comparison of the transcription factor coexpression pattern.

The involvement of the mushroom bodies in place learning has been demonstrated in cockroaches (Mizunami et al., 1998). Cockroaches with mushroom body ablations between the beta lobe and pedunculus were not able to learn the place of a hidden target by using distant visual cues (Mizunami et al., 1998). It was also shown that mushroom body extrinsic neurons are activated during specific behavioral sequences in freely moving cockroaches (Mizunami et al., 1998). For the honeybee and *Drosophila* the mushroom body, a neuropile which consitsts of intrinsic

Kenyon cells (Kenyon, 1896), is necessary for olfactory learning and memory as was shown in learning studies in *Drosophila* mushroom body mutants (Heisenberg et al., 1985) and mushroom body cooling experiments in the honeybee (Masuhr, 1976, Erber et al., 1980). The *Drosophila* mushroom body mutants were still able to learn a color discrimination task (Heisenberg et al., 1985). In contrast, mushroom body mutations in the fly brain do not affect visual operant learning (Wolf et al., 1998). Nevertheless, context generalization in visual learning in *Drosophila* requires intact mushroom bodies (Liu et al., 1999). The applies also for locomotion control (Martin et al., 1998). It was shown that decision making is partly dependent on the mushroom body (Zhang et al., 2007). During simple linear perceptual decision making the mushroom bodies are not required (Zhang et al., 2007). *Drosophila* mushroom body mutants show this linear form of decision making but not the non-linear form which is seen in saliency based situations in wild types, deciding between conflicting cues (Zhang et al., 2007).

The recorded mushroom body extrinsic neurons are known to change their properties during olfactory learning, sometimes after a consolidation phase of several hours (Okada et al., 2007, Strube-Bloss et al., 2011). These neurons receive input from the intrinsic neurons of the mushroom body. They are sensitive to combinations of multiple sensory modalities including visual stimuli (e.g. the PE1 neuron (Mauelshagen, 1993, Rybak et al., 1998)). The Vummx1, a neuron which is encoding a reward related prediction error (Hammer, 1993) and the PE1 are projecting to the lateral horn (Mauelshagen, 1993). The lateral horn neuropile is innervated by descending sensory-pre-motor connections in the cockroach (Okada et al., 2003). Therefore it seems likely that the PE1, a unique neuron in the ventral alpha lobe, is involved in the initiation of motor output on the basis of reward predictions. This is a prerequsite for decision making and operant learning.

Rats and humans are able to navigate in visual virtual environments (Mallot et al., 1998, Höllscher et al., 2005). Neural recordings from the hippocampus are possible during functional neural imaging (Dombeck et al., 2010) and even from single neurons intracellularly (Harvey et al., 2009). *Drosophila* flying in a simple virtual environment has helped to elucidate a large range of visual performances and visual learning at multiple levels of analysis since more than 30 years (Wolf et al., 1991, Peng et al., 2007) . However, combining flight behavior in a virtual environment with neural recordings has turned out to be rather difficult in *Drosophila* leading to some correlations between turning behavior and local field potentials (Van Swinderen, 2003). Intra- and extracellular recordings during olfactory learning in restrained honeybees helped to understand multiple facets of sensory encoding and neural correlates of memory formation (Menzel et al., 2006, Denker et al., 2010, Strube-Bloss et al., 2011) but instrumental forms of learning, related to visual navigation was not possible so far in honeybees. Here we used an air-supported spherical treadmill that allows the stationary walking honeybee to control the visual environment while long lasting extracellular multi unit recordings were performed from mushroom body extrinsic neurons. Since the visual system of honeybees (and other insects) differs in many aspects from that of humans, it is a rather difficult task to create an immersive VE for a honeybee. For example bees are able to detect UV light and polarized light (von Frisch (1914), Lindauer (1959), Menzel (1975)). Therefore it is important to compare the behavioral and electrophysiological studies of decision making in a virtual environment with comparable measurements in a real setting. To record from freely walking honeybees, a chronically implantable plug for extracellular long term recordings in honeybees (utility patent Nr. 20 2012 002 773.5, de Camp, http://register.dpma.de) was developed.

Additionally the plug enables a comparison of electrophysiological studies in freely walking honeybees with well-established paradigms in vertebrates like rats and mice (Hussaini (2011), von Heimendahl (2012)).

Material and Methods

Virtual Environment Setup. Spherical treadmill, geometry of the virtual environment and overall setup were described by Höllscher et al. (2005), apart from the following changes (see also Fig. M1): The treadmill consisted of a Styrofoam ball (10 cm diameter) placed in a half-spherical plastic cup with several, symmetric located holes through which a laminar air flow passed and let the ball float on air. Laminarity of the air flow was supported by a long (12 m) tube from the well regulated fan to the ball. Low static electrification of the ball and humidity for the animal was achieved by blowing the air into a box with water. Buoys prohibited corrugation inside the box.

The beamer Epson EMP-TW 700 (digital scanning frequency: pixel clock: 13,5-81 MHz, horizontal sweep: 15-60 kHz, vertical sweep: 50-85 Hz) was positioned above the Faraday cage and illuminated the inner surface of a cone shaped screen (height 60 cm, bottom diameter 7 cm, top diameter 75 cm) via a large surface mirror and a Perspex window (Fig.1). A Fourier power analysis of the beamer projection via a photodiode (thanks to Uwe Greggers) revealed a main part 210 Hz frame rate. Less than 10% 100Hz and 60Hz have been measured, respectively (measurement 15.2.2010, kindly provided by Uwe Greggers). The inner surface of the cone consisted of white paper. Patterns projected onto this screen were distorted such that they appeared undistorted when seen by the bee (BeeWorld, programmed by Sören Hantke).

During an experiment, the Faraday cage was closed. A webcam (Logitech, Morges Gesellschaft) positioned above imaged the head of the animal via a 500 mm mirror objective allowing an observation of the animal during the experiment.

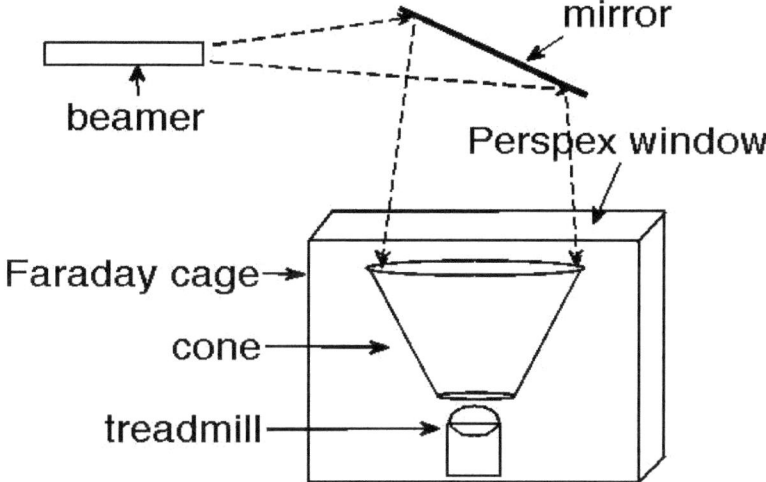

Fig. M1: The light was projected from a beamer onto a mirror and from the mirror onto the inner surface of a cone shaped sreen. The screen was paced inside a Faraday cage with a Perspex window at the top, to allow for a passage of the light. The screen could be lifted to place the animal on the styrofoam ball (treadmill).

Control of the virtual environment and experimental procedure. The virtual environment and the recognition of the hornets movement was under the control of the customs written program BeeWorld (Sören Hantke). It was implemented in Java by using OpenGL-Bindings for Java (LWJGL). The movement of the ball, initiated by the walking animal, was detected by two optical computer mice as they are used for computer games (Imperator, Razer Europe GmbH; G500, Logitech Europe S.A.). The mice were accurately positioned under 90° at the equator of the Styrofoam ball and precisely aligned with x/y micro drives. The animal was able to control the virtual scenery by rotatory movements of the ball. Translatory movements magnified the objects on the screen as if the animal would approach. Multiple scenarios were programmed. They were realized as xml-files containing the positions, width and color (RGB) of a variable number of vertically oriented stripes. These stripes were positioned around the hornet. Scenario7 consisted of alternating black and white, vertically oriented stripes with an angular width of 20° and an grey interspace of 40°. In the figures presented here, the redundant pattern is standardized on one black and one white pattern in an angle of 120°. Therefore, 5° in the plots are equvalent to 15° in the virtual environment. Scenario12 consisted of one black, one white and four blue stripes with the same dimensions as in scenario7.

The field of view of a camera in OpenGL is limited to 179°, the scenarios projected onto the screen, however, simulated a 360° world. In the BeeWorld program a four texture renderer (texturerenderer) was used to create a 360° camera. Data of walking traces were synchronized with the data from spike recordings which were collected with an analog to digital converter (micro3, CED, Cambridge Electronic Design, 30 kHz sampling frequency per channel). A Silicon NPN Phototransistor (BPY 62) directed at the screen detected a short light signal under the control of the BeeWorld program and fed it into the ADC input of the analog to digital converter. The pulse was recorded in a spike2 channel and allowed for precise timing between the spike and walking data, recorded with different computers.

A skyline in the background of the scenario consisted of vertically oriented stripes of different hues of grey and different length. A checker board pattern was projected on the plain floor around the hornet. The angular rotation of this checker board pattern was equal to the angular movement of the ball simulating a respective movement of the floor directly below and around the animal. The angular rotation of the stripe pattern was set to 75% of the checker board pattern, and a skyline pattern projected onto the screen together with the stripe pattern moved with 50% of the checkerboard pattern. Thus these three patterns simulated depth information by creating a signal of different motion parallax as seen by the stationary walking insect.

Electrophysiology in stationary walking *Hymenoptera*. Four insulated copper wires (Elektrisola) with a diameter of 0.015mm each were coiled and fixed with superglue or heated (210° for 3 minutes have been appropriate to fuse the polyurethane insulation of the wires without producing an electric shortage). These techniques increased the stability of the coiled electrode bundles and facilitated tissue penetration.

The best signal to noise ratio was achieved with Teflon insulated silver ground wires with a diameter of 0,125 mm (WPI, Berlin, D) implanted into the abdomen or Teflon insulated Platinum/Iridium wires of 0.025mm diameter (advent, Enysham, Oxford, UK) which were thin enough to be implanted in the brain, compound eye or ocelli. The insulation at the tip of these wires was removed mechanically with a fine forceps.

The tip of the tetrode was cut with an fine scissors. The single ends of the copper wires were dipped into hot solder to remove the insulation and then connected with silver conductive paint (electrolube) to the female side of the pins of an IC socket. After drying, the coiled ends of the electrodes were electroplated using the low-impedance plating procedure with gold and PEG (Ferguson et al., 2009) with the electroplating device NanoZ (Neuralynx). A scanning electron microscope picture of such a partially plated tetrode was kindly provided by Prof. H. Hilger (Fig. M2).

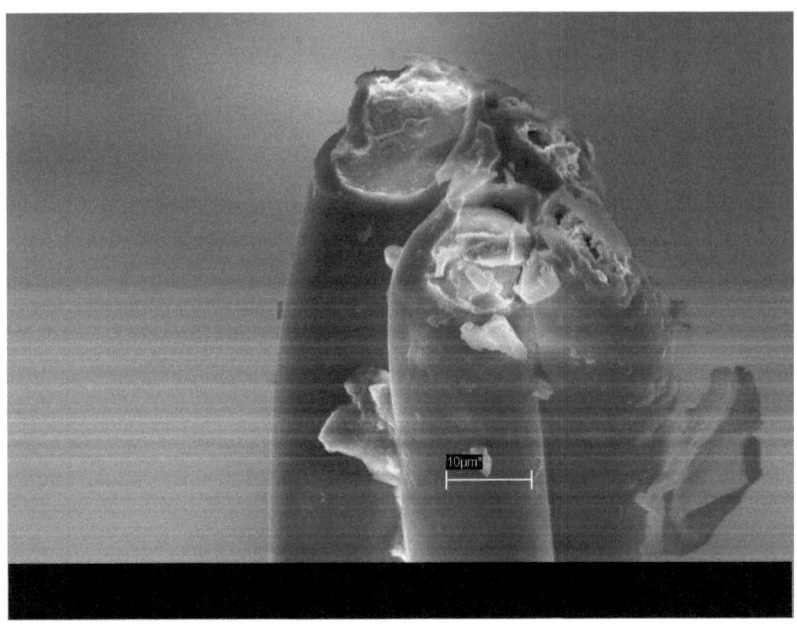

Fig. M2: **A scanning electron microscopy picture of a gold plated tetrode (kindly provided by H. Hilger).** The scale bar indicates 10µm. The PU insulated copper wires were stuck together by heat. The plating procedure (Ferguson et al., 2009) results in low-impedance electrodes via increased surface area. The increase of surface is due to the textured accumulation of gold and PEG particles on the copper surface.

Preparation of the animals. The Bee was caught at the hive entrance and the hornet from a flight cage. The following descriptions are the same for the bee and the hornet. The animals were chilled on ice and fixed temporarily in a small tube with modelling clay. A small piece of plastic tube or rubber foam was fixed with dental wax on the thorax as holder for the stationary running animal on the treadmill. A window between the compound eyes, antennae and ocelli was cut into the head capsule. The tip of the electrode was fixed to a fine forceps which was mounted on an external micromanipulator. The electrode was inserted into the brain, while the animal was still harnessed in the tube. After placing the electrodes in the selected brain area (ventral aspect of the alpha lobe of the mushroom body) under visual control, the electrode was fixed with two component silicone elastomer (kwik sil, WPI) onto the brain and the head capsule. After hardening of the kwik sil, the electrode was released from the external micromanipulator and additionally fixed inside a small slit in the plastic tube/rubber foam holder on the thorax by forming a small loop from the head backwards to the thorax. About 5 minutes later, the bee was released from the tube by grabbing the plastic tube/rubber foam mounted onto the thorax with a forceps and pushing the head with another forceps slowly backwards to facilitate the animals release from the tube. The rubber foam had a slit and the electrodes could be fixed without silicone elastomer. Fixation of the electrodes, however, was crucial for stable and long lasting recordings. Afterwards the animal was clipped by the plastic tube or piece of rubber foam on its thorax to an alligator clip attached to the electrode holder. Longer electrodes were additionally fixed with modelling clay to the electrode holder to prevent reachability of the fine wires from the range of the bee's legs. Especially during the transfer to the setup after the release from the fixation tube and during the first minutes on the ball the animals often tried to remove the electrodes. The electrode holder with the hornet was rotated 90° and the animal was carefully adjusted onto the floating ball. The electrode holder consisted of a small balance which kept the animal on the treadmill with its own weight. This balance allowed the animal also to change the distance to the surface of the treadmill during walking. The direct light from the LCD projector was shaded in order to prevent direct illumination of the dorsal regions of the compound eyes and the ocelli. Two UV diodes were positioned within the shade just above the head of the animal simulating short wavelength light coming from above. The micromanipulator and binoculars necessary for positioning the tetrode during the preparation procedure were removed from the setup when the experiment started. Another manipulator allowed precise positioning of the animal on the spherical treadmill.

Extracellular recordings with chronically implanted electrodes. The functional core of the newly developed micro plugs for chronic implantation of extracellular electrodes into the brains of small insects of approximately honeybee size are at least two pieces of plastic tubes (Portex polythene tubing, Smiths Medical, Kent, UK, inner diameter (id) 0.28mm, outer diameter (od) 0.61mm, length (depending on the animal) for honeybees about 1.5mm). The polythene tubes inner surface was covered with a conducting fluid by capillary forces (silver conductive paint, Electrolube or Graphit 33, CRC Industries Europe NV). After drying, the quality of the conductive cover on the inner surface was controlled visually.

To construct the smallest plug for one differential extracellular recording channel, three pieces of polythene tubing with a conductive inner surface were needed. Two copper wires (18μm diameter, polyurethane insulation, Elektrisola) were coiled as described above. The coiled wires were cut into pieces of approximately 2 cm length. One end of these small pieces of coiled wires were uncoiled carefully with a fine forceps. The polyurethane insulation of these uncoiled ends was removed by heat (solderer, 400°C). Each single end of the coiled copper wire was inserted in one piece of polythene tubing with inner coating and fixed with a silver glue. For appropriate grounding it was important to have a large metal surface in tissue contact. Therefore a Pt90/Ir10 wire with quadruple Teflon insulation was used (id 0.025mm, od 0.035mm, Advent). The insulation was mechanically removed with a forceps at both ends of the PtIr wire. The wire was fixed in a piece of polythene tubing as described for the copper wires. The ends of the tubes with the inserted wires were insulated with superglue. Thereafter the tubes were fixed as a bundle with superglue in such a way, that the open ends of the two recording wires (copper) showed to the same direction and the PtIr wire (ground) open end in the opposite direction. A 2x2x2mm piece of foam rubber was used as bearing for the tubes and wires. The foam rubber was trimmed such that the tubes were sunk in a groove and the other side of the foam had a concave form, oppositely to the convex form of the bees thorax (for a perfect fit when the plug was fixed on top of the bees thorax). From this bee attachment side of the foam, it was penetrated with a fine forceps and the three wires were pulled through the foam. This procedure was necessary to attach the wires near to the bee's body. Otherwise the bees were able to destroy the thin wires with leg movements. The conglomerate of tubes and wires was fixed onto the piece of foam rubber with super glue and the whole plug was fixed with dental wax onto the bees thorax. The PtIr ground wire was implanted in one eye or into the abdomen. Careful fixation of the wires onto the head capsule with silicone elastomer (kwik sil, WPI, Sarasota) was important for long lasting recordings without movement artefacts.

The second plug, for the connection of the chronically implanted electrodes with the preamplifiers, was built by an IC socket with three long copper wires (40cm length, 18µm diameter, polyurethane insulation, Elektrisola).

The ends of the long copper wires were dismantled with a solderer (as described previously) and each one fixed with silver conductive paint and hard wax to a 1cm long piece of Teflon insulated silver wire (125µm diameter) which was mechanically dismantled at both ends beforehand. The three resulting silver wire ends were fixed to each other with silicone elastomer (kwik sil, WPI). The bare ends were formed like a bind rail with two wires arcuated in one direction and one (the ground connection) to the opposite direction in a way that the three ends geared into each other. In combination with the elastic silicone component this construction guaranteed a stable fitting into the polythene tubes of the chronically implanted electrode device on top of the insect thorax without administering vertical pressure onto the animal. The recordings were performed in a 10x10 cm arena (Fig. R19). This arena was subdivided into 4 quadrants (Q) for analysis. Q1 comprised a sugar source on top of a blue piece of paper (the main ground of the arena was bee wax). Q2 comprised water on a yellow paper, Q3 was empty and Q4 comprised a small tube with beeboost (Queen Pheromone with Queen Mandibular Pheromone, Contech Enterprises Inc., Canada).

Analysis. Spike2 (Cambridge Electronic Design) Software was used for data sampling and spike sorting. Before sorting, the data were high pass filtered. For ripple detection, the data were low pass filtered. Matlab (2010b, MathWorks) was used for statistical analysis and the coefficient of correlation (Spearman's Rho). The data presented here were tested on normal distribution (Kolmogorov-Smirnov test, lilliefors test), with a negative result. Therefore, non parametric tests were used, like the Wilcoxon ranksum test to compare two variables or the Kruskal-Wallis test to compare more than two groups. The mean plus standard deviations are shown for the spike rate data. Due to the high sample size in these cases, the mean plus standard deviation was a valid method to approximate that 68% of the data are in the interval of mean plus one time standard deviation, 95% of the data are found in the interval of mean plus two times standard deviations and 99% of the data are found in the interval of mean plus three times standard deviation. The place fields are calculated for unit activity other than during the high frequency ripple associated phases. A ripple correction factor was calculated, to test the influence of ripple events to the place fields (Fig.R18). Therefore, the mean ripple length was multiplied with the increase in frequency in comparison to average non-ripple spiking of unit1. The resulting correction factor gives the mean number of spikes which are added by one ripple event. This factor was multiplied to a vector of ripple number/5° bins of the virtual space and subtracted from the calculated unit1 firing rate.

To correct for high firing rates at a certain place due to a longer duration of stay and an increased probability for the occurrence of high phases, the count of spikes per virtual spatial bin was divided by the time, spent in this part of the virtual space.

Results. Ripple like potentials have been observed in a freely moving honeybee and a virtually navigating hornet. The ripples are characterized by an on phase, a positive shift in the baseline potential of the recording channel and an off phase, a negative shift (both together are defined as ripple cycle). Furthermore, both recordings have a strong increase in unit activity during the ripple cycles in common. Not all recorded units show an increase during the ripple events. From 3 recording channels in the case of the virtually navigating hornet, the ripples can only be observed in one channel. During navigation, the ripples occur most often during immobile wakefulness or during sleep (in the freely moving bee). But the ripples are not restricted to these behavioral states and are also observed during locomotion activity. The units which are associated with increased firing rates during the ripple-like events (for simplicity called ripples), show place related firing activity in the virtual space (Fig. R 17).

Results for the virtually navigating hornet. The hornet had rotatory and translatory control during virtual navigation. The animal was recorded for four consecutive days in the ventral border region of the mushroom body alpha lobe. The ripple like events are starting already in a test phase, when the animal was exposed to the floating ball and the setup for the first time – without visual feedback. As shown in figures R1 to R6, the ripple events are correlated with increased unit1 frequency. The high frequency phases of unit 1 are enframed by ripple events (fig. R1).

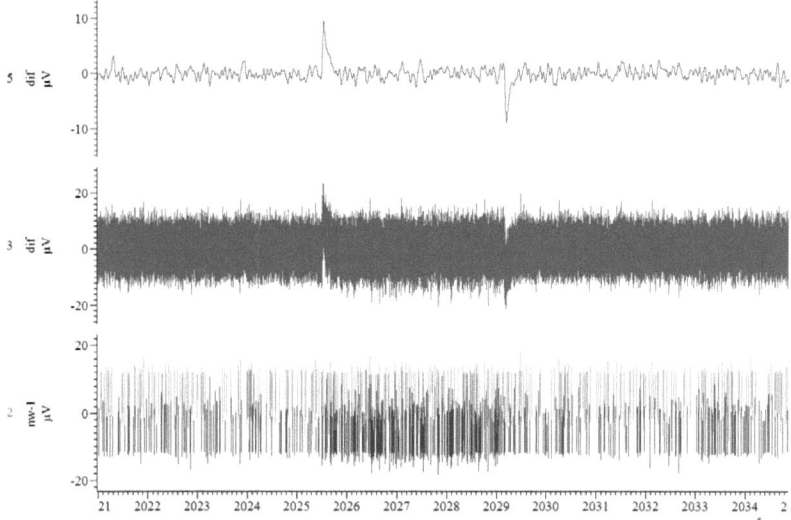

Fig. R1: Illustration of ripple cycles and associated unit activity in the original recording trace. Channel 3 (trace in the middle) is the original recording channel in which the ripple events occured. The ripples are appearing as pairs of one positive and a following negative baseline deflection, which are enveloping high unit activity, as seen in channel 2 (lower trace, u1 in blue and u2 in green). To extract the ripple events from the recording channel, channel 3 has been downsampled and low pass filtered. The result is seen in channel 5 (top row).

A transient positive baseline deflection in the extracellular recording and a following negative deflection are enframing the high unit 1 activity. The duration of such ripple cycles is varying (fig. R4, positive ripple (+), negative ripple (-)). Ripple cycles (measured as inter ripple interval from the positive to the following negative ripple event) of less than one second are occuring as well as cycles with 10s and even more (fig. R4). The ripple waves are characterized by a very steep onset (approximately 0.01s) and a slow after-peak-decline, which leads to a total time of approximately 0.2s per ripple (regardless if positive or negative). The ripples are occuring most likely in pauses during navigation (fig. R2-R6, third row in fig. R2-R6, respectively). These phases are charcterized by low rotation acivity (the first row in fig.R2-R6, respectively) and low translation (forward running) acitivity (the second row in fig. R2-R6, respectively). The correlation between unit 1 activity and ripple is evident during all 4 days of the experiment (fig. R2-R6, third row: ripples, fourth row: unit 1 frequency). As seen in figures R5 and R6, unit 2 frequency (fifth row, respectively) is sometimes correlated with the ripple cycles (row three, respectively), but not as clear as for unit 1 (row four, respectively). Unit 2 is recorded with the same elctrode as unit 1 and the ripples. In contrast, unit 3 is recorded with another electrode (in the same region of the mushroom body alpha lobe). Unit 3 is correlated with the walking activity in the virtual environment at the second day of the experiment (fig. R2, bottom row). This correlation can not be found at the fourth day (fig. R3, R6, bottom row, respectively). The animal was fed with sucrose solution (50%) between the recording sessions. During feeding, a blue stripe of paper was presented in front of the animal. At the fourth day, the hornet was tested in szenario 12 with four blue stripes (for a detailed description see materials and methods) (fig. R4-R6). The ripples are also occuring in this new context (fig. R4-R6, third row, respectively). Interestingly, the activity of unit 2 is stronger correlated with the ripple cycles in this secenario with blue stripes (fig. R4, R5, fifth row, respectively) than in scenario 7 with white and black stripes (fig. R2, R3). Unit 3 is active during virtual navigation in scenario 12 with vertically oriented blue stripes, but no obvious behavioral correlation can be seen (fig. R4, bottom row).

Until day 3, two other units are occuring, which are not correlated with the ripple events (Fig. R7). Unit 4 is correlated with the walking activity of the animal.

The walking pattern was analyzed for the single components duration of stay, translatory activity and translation velocitiy in different VEs. At the second day, there is no obvious difference between the control situation without visual feedback (Fig. R8-10a) and the duration of stay (Fig. R8b), translation (Fig. R9b) and forward running speed (R10b) in the virtual szenario 7 with alternating vertical black and white stripes. At the fourth day, the duration of stay (Fig. R11a, b) is high for regions apart from cues (Fig. R11b) in szenario 7 with black and white stripes in comparison to a control situation (Fig. R11a), without visual feedback for the animal. Additionally, the translation (Fig. R12) and running speed (Fig. R13) are increased towards the white stripe in szenario 7 (Fig. R12b, 13b) in comparison to the control situation without visual feedback (Fig. R12a, 13a). During the test situation in szenario12 with blue stripes, the animal shows highly consistent preference for the white stripe and the blue stripes during all three conditions (duration of stay: Fig. R14a,b; translation: Fig. R15a,b; speed: Fig. R16a, b). In this virtual szenario the animal shows no running behavior towards the black stripe.

Unit 1 and 2 but not unit 3 show place related firing (Fig. R17), which is independent of the occurence of ripple events (ripple correction in Fig. R18) and the duration of stay (normalized firing rates in figures R17 and R18, both correction/normalization procedures are described in the methods). The 3D relation of place realted firing patterns is shown in figure R19. Spiking activity during the virtual trajectory is coded in cyan. The unit

1 activity is increased in the beginning of the trajectory, especially during a navigation pause.

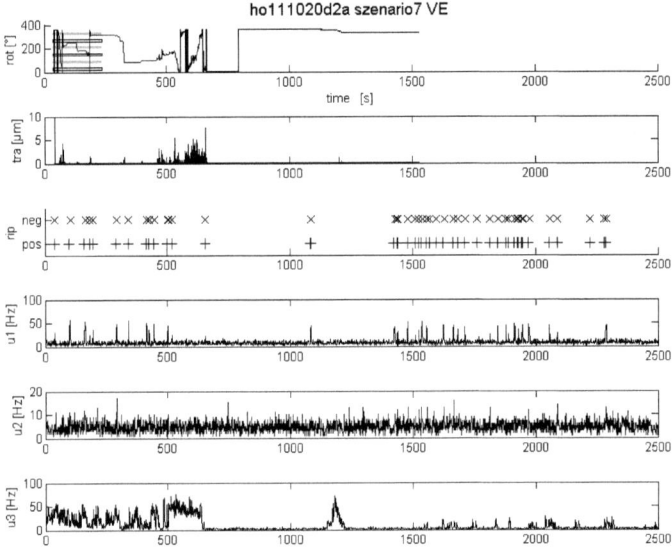

Fig. R2: Rotation, translation and spike activity at experimental day 2 in a virtually navigating hornet. The first row (rot on the Y-axis) shows the rotatory activity in Szenario7. The layout of the szenario is indicated on the left side, alternating black (plotted in grey) and white stripes (plotted with black edges). The black line is the hornets rotation trajectory. ...

...FigR2: ... Below (Y-axis tra), the translatory activity is plotted. The instantaneous translation was calculated as translation difference between two mouse update events (approximately 25ms). The Y-axis rip (third row) shows the ripple events. Pos is the onset of one ripple cycle, characterized by a positive baseline deflection and neg means the end of one ripple cycle, with a negative deflection. The ripple cycles are highly correlated with u1 (fourth row) activity but not u2 activity (fifth row) on the same recording channel. U3 frequency (sixth row) is not correlated with the ripple events but with the walking activity of the animal.

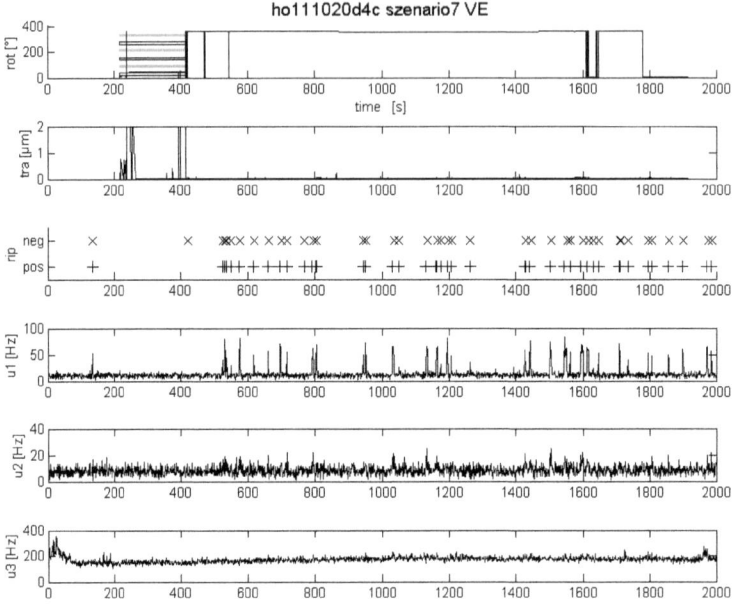

Fig. R3: Rotation, translation and spike activity at experimental day 4 in the virtually navigating hornet. Again, the first row (rot on the Y-axis) shows the rotatory activity in Szenario7. The layout of the szenario is indicated on the left side, vertically oriented, alternating black (plotted in grey) and white stripes (plotted with black edges). The black line is the hornets rotation trajectory. Below (Y-axis tra), the translatory activity is plotted. The instantaneous translation was calculated as translation difference between two mouse update events (approximately 25ms). ...

...FigR3: ... The Y-axis rip (third row) shows the ripple events. Pos is the onset of one ripple cycle, characterized by a positive baseline deflection and neg means the end, with a negative deflection. The ripple events are also at day 4 highly correlated with u1 (fourth row) activity and less strong u2 activity (fifth row) on the same recording channel. The frequency of unit 3 (bottom row) is neither correlated with the occurence of ripple events nor the animals walking activity in contrast to the second day of recording.

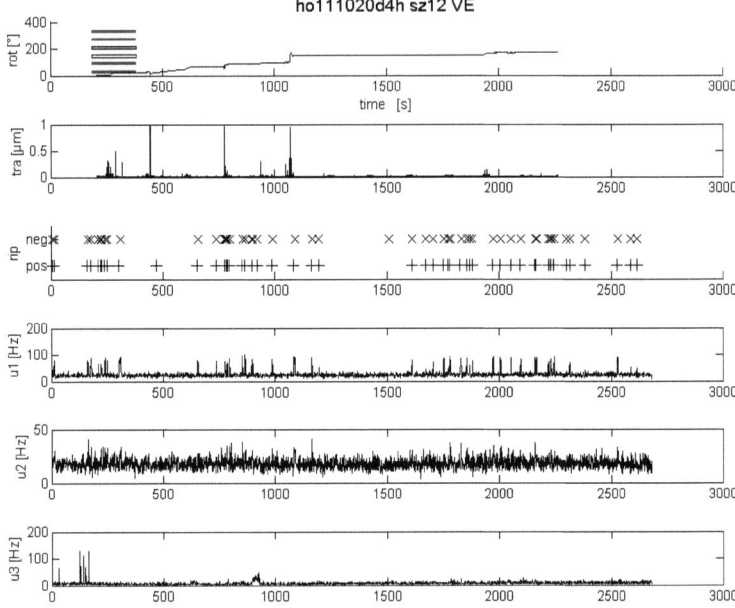

Fig. R4: Rotation, translation and spike activity at experimental day 4 in the virtually navigating hornet. The first row (rot on the Y-axis) shows the rotatory activity in Szenario12. The layout of the szenario is indicated on the left side. From the perspective of the hornet the stripes are vertically oriented. The virtual environment consists of one black (plotted in grey), one white and four blue stripes. The black line is the bees rotation trajectory. Below (Y-axis tra), the translatory activity is plotted. The instantaneous translation was calculated as translation difference between two mouse update events (approximately 25ms). ...

...FigR4: ... The Y-axis rip (third row) shows the ripple events. Pos is the onset of one ripple cycle, characterized ba a positive baseline deflection (+) and neg means the end of one ripple cycle, with a negative deflection (-). The ripple events are also during navigation in this newly presented virtual szene highly correlated with u1 activity (fourth row). Unit 2 activity (fifth row) on the same recording channel is less correlated with the ripple cycles than unit 1 frequency. Again, the frequency of unit 3 (bottom row) is neither correlated with the occurence of ripple events nor the animals walking activity.

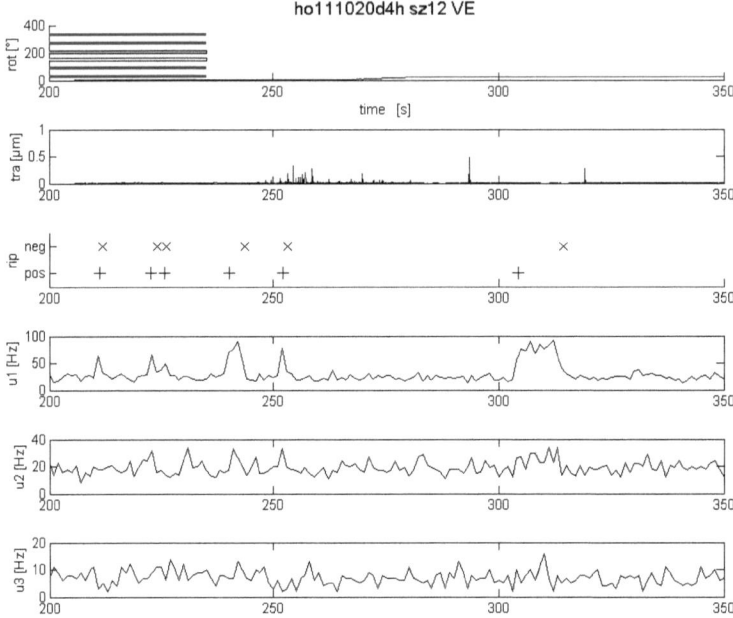

Fig. R5: Detail of fig. R3. Rotation, translation and spike activity at experimental day 4 in the virtually navigating hornet. The first row (rot on the Y-axis) shows the rotatory activity in Szenario12. The layout of the szenario is indicated on the left side. From the perspective of the hornet the stripes are vertically oriented. The virtual environment consists of one black (plotted in grey), one white and four blue stripes. The black line is the bees rotation trajectory. Below (Y-axis tra), the translatory activity is plotted. The instantaneous translation was calculated as translation difference between two mouse update events (approximately 25ms). The Y-axis rip (third row) indicates the ripple events. ...

...FigR5: ... Pos is the onset of one ripple cycle, characterized ba a positive baseline deflection (+) and neg means the end of one ripple cycle, with a negative deflection (-). The ripples are highly correlated with unit 1(fourth row) activity. This detail from fig. R3 shows ripples during navigation in near vicinity to blue stripes. Unit 2 (fifth row) is less clear correlated with the ripple cycles. The frequency of unit 3 (bottom row) is neither correlated with the occurence of ripple events nor the animals walking activity.

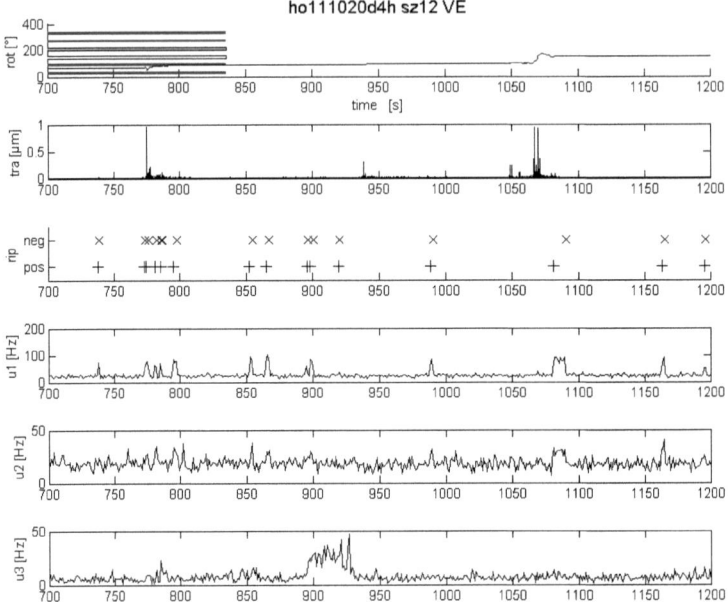

Fig. R6: Detail of fig. R3. Rotation, translation and spike activity at experimental day 4 in the virtually navigating hornet. The first row (rot on the Y-axis) shows the rotatory activity in Szenario12. The layout of the szenario is indicated on the left side. From the perspective of the hornet the stripes are vertically oriented. The virtual environment consists of one black (plotted in grey), one white and four blue stripes. The black line is the bees rotation trajectory. ...

...FigR6: ... Below (Y-axis tra), the translatory activity is plotted. The instantaneous translation was calculated as translation difference between two mouse update events (approximately 25ms). The Y-axis rip (third row) shows the ripple events. Pos is the onset of one ripple cycle, characterized ba a positive baseline deflection (+) and neg means the end of one ripple cycle, with a negative deflection (-). The frequency of unit 3 (bottom row) is neither correlated with the occurence of ripple events nor the animals walking activity.

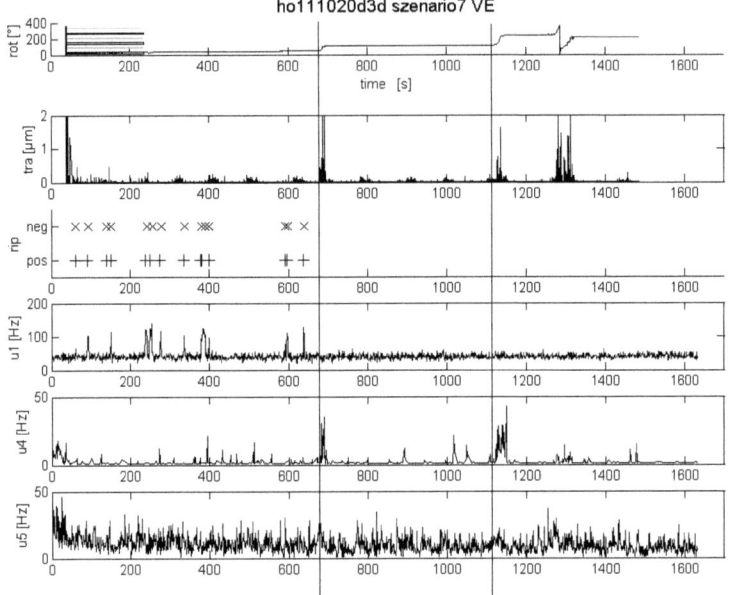

Fig. R7: Unit 4 and 5 activity. Rotation, translation and spike activity at experimental day 3 in the virtually navigating hornet. The first row (rot on the Y-axis) shows the rotatory activity in Szenario7. The layout of the szenario is indicated on the left side. From the perspective of the hornet the stripes are vertically oriented. The virtual environment consists of alternating black (n=3, plotted in grey) and white stripes (n=3). The black line is the hornets rotation trajectory. Below (Y-axis tra), the translatory activity is plotted. ...

...FigR7: ... The instantaneous translation was calculated as translation difference between two mouse update events (approximately 25ms). The Y-axis rip (third row) shows the ripple events. Pos is the onset of one ripple cycle, characterized ba a positive baseline deflection (+) and neg means the end of one ripple cycle, with a negative deflection (-). **Unit 4 (fifth row) seems to be correlated with the animals walking activity (note the activity near to the vertical lines). The frequency of unit 5 (bottom row) is neither correlated with the occurence of ripple events nor the animals walking activity.**

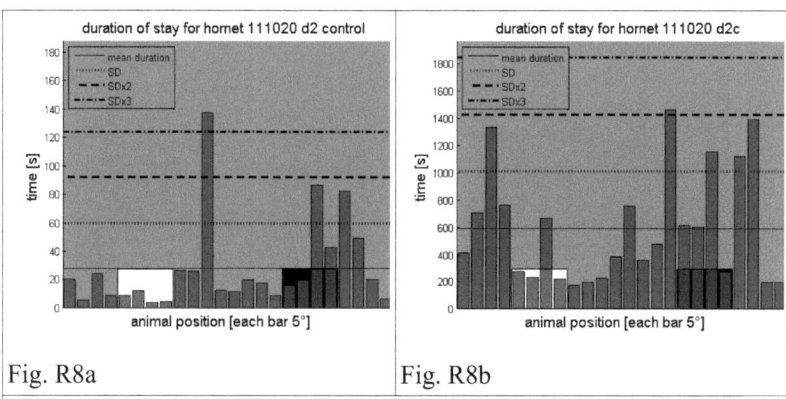

Fig. R8a   Fig. R8b

Fig. R8: Duration of stay for hornet 111020 at the second day in the virtual environment. The white and light grey bar indicate the position of the white and black vertically presented stripes in the virtual szene. The whole szene consisted of 6 alternating black and white stripes, each with a widh of 20°. The slim bars represent the duration of stay for the animal in the virtual space. Since the duration of stay is normalized to one black and one white stripe (120° of the VE), each bar represents 5°x3=15° of the full panorama. The horizontal lines indicate mean duration of stay and the mean plus 1,2 and 3 times standard deviation (SD). The higher amount of duration of stay at the borders of the figure is partly due to the starting point of the animal in the VE. The control situation (Fig.R8a) is without any visual feedback. **At day 2c (Fig. R8b) the animal is preferring the border regions of the black stripe.**

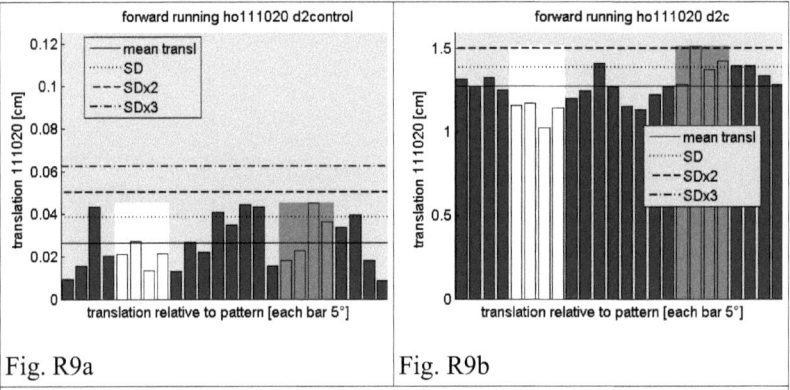

Fig. R9a   Fig. R9b

Fig. R9: Forward running (translation) for hornet 111020 at the second day in the virtual environment. The white and light grey bar indicate the position of the white and black vertically presented stripes in the virtual szene Nr.7, which was presented here. The whole szene consisted of 6 alternating black and white stripes, each with a widh of 20°. The slim bars represent the duration of stay for the animal in the virtual space. Since the duration of stay is normalized to one black and one white stripe (120° of the VE), each bar represents 5°x3=15° of the whole panorama. The horizontal lines indicate mean translation and the mean plus 1,2 and 3 times standard deviation (SD). The higher amount of duration of stay at the borders of the figure is partly due to the starting point of the animal in the VE. **Please note the equal distribution of translatory activity with** visual feedback (fig. R9b) in comparison to the control (fig. R9a) without visual feedback.

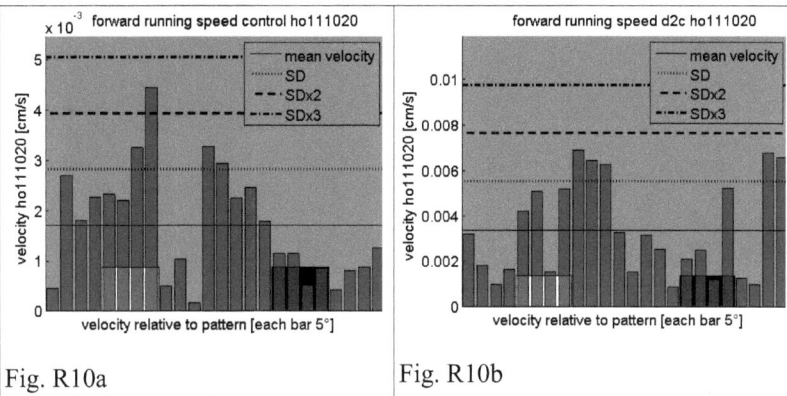

Fig. R10a   Fig. R10b

Fig. 10: Forward running (translation) velocity for hornet 111020 at the second day in the virtual environment. The white and light grey bar indicate the position of the white and black vertically oriented stripes in the szene Nr. 7, which was presented here. It consisted of 6 alternating black and white stripes, each with a widh of 20°. The slim bars represent the running speed for the animal in 5° of the virtual space, respectively. Since the duration of stay is normalized to one black and one white stripe (120° of the whole VE), each bar represents 5°x3=15° of the panorama. The horizontal lines indicate mean translation velocity and mean velocity plus 1, 2 and 3 times standard deviation (SD). **There are no obvious differences in the spatial distribution of running speed at the second experimental day 2c (Fig. R10b) and the control (Fig. R10a).**

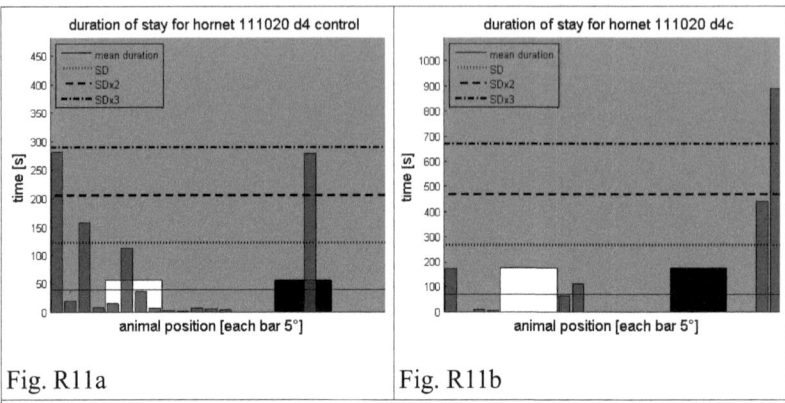

Fig. R11a　　　　　　　　　　　Fig. R11b

Fig. R11: Duration of stay for hornet 111020 at the fourth day in the virtual environment. The white and light grey bar indicate the position of the white and black vertically presented stripes in the virtual szene. The whole szene consisted of 6 alternating black and white stripes, each with a widh of 20°. The slim bars represent the duration of stay for the animal in the virtual space. Since the duration of stay is normalized to one black and one white stripe (120° of the VE), each bar represents 5°x3=15° of the whole panorama. The horizontal lines indicate mean duration of stay and the mean plus 1,2 and 3 times standard deviation (SD). The higher amount of duration of stay at the borders of the figure is partly due to the starting point of the animal in the VE. In the Control situation (fig. R11a) no stripe pattern was presented. **Please note the higher duration of stay in the area of the black vertical stripe in the control situation without any visual feedback. At day 4c the animal is not walking on the stripe areas as in the control of the second day or at day 2c.**

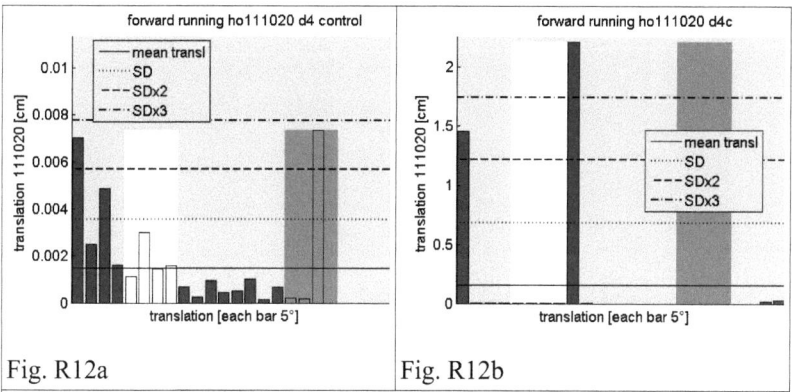

Fig. R12a Fig. R12b

Fig. R12: Forward running (translation) for hornet 111020 at the fourth day in the virtual environment. The white and light grey bar indicate the position of the white and black vertically presented stripes in the virtual szene Nr.7, which was presented here. The whole szene consisted of 6 alternating black and white stripes, each with a widh of 20°. The slim bars represent the duration of stay for the animal in the virtual space. Since the duration of stay is normalized to one black and one white stripe (120° of the VE), each bar represents 5°x3=15° of the whole panorama. The horizontal lines indicate mean translation and mean plus 1,2 and 3 times standard deviation (SD). The higher amount of duration of stay at the borders of the figure is partly due to the starting point of the animal in the VE. In the Control situation (fig.R12a) no stripe pattern was presented. **Please note the high amount of translation at day 4c (Fig. 12b) in contrast to the control situation without visual feedback (Fig. R12a).**

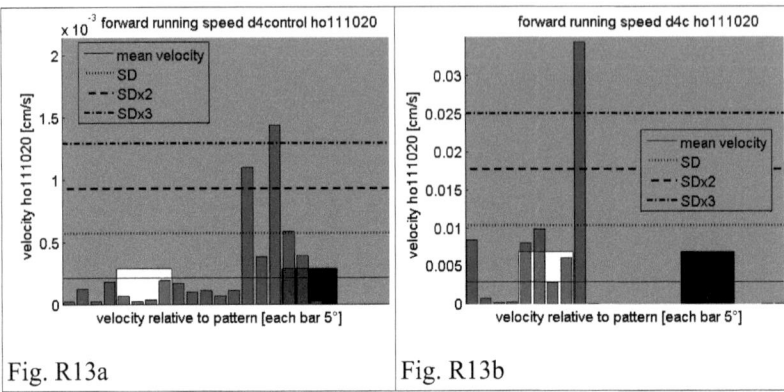

Fig. R13a   Fig. R13b

Fig. R13: Forward running (translation) velocity for hornet 111020 at the fourth day in the virtual environment. The white and black bar indicate the position of the white and black vertically oriented stripes in the szene Nr. 7, which was presented here. It consisted of 6 alternating black and white stripes, each with a widh of 20°. The slim bars represent the running speed for the animal in 5° of the virtual space, respectively. Since the duration of stay is normalized to one black and one white stripe (120° of the whole VE), each bar represents 5°x3=15° of the panorama. The horizontal lines indicate mean translation velocity and mean velocity plus1, 2 and 3 times standard deviation (SD). **The forward running speed is maximal towards the white stripe (Fig. R13b) in comparison to a control situation without visual feedback (Fig. R13a).**

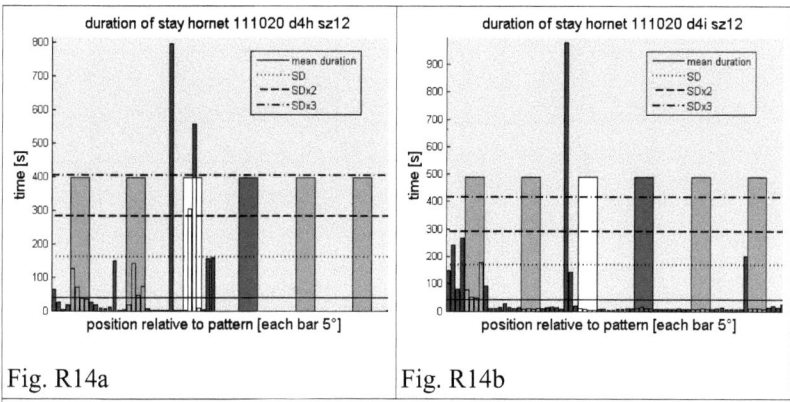

Fig. R14a   Fig. R14b

Fig. R14: Duration of stay for hornet 111020 at the fourth day in the virtual environment. The white and dark grey bar indicate the position of the white and black vertically presented stripes in the virtual szene 12. The light grey bars indicate blue stripes. Each stripe has a width of 20°, the x-axis represents the 360° virtual panorama. The slim bars indicate the duration of stay for the animal in the virtual space (each bar 5°). The horizontal lines indicate mean duration of stay and mean plus 1, 2 and 3 times standard deviation (SD). **Please note the higher duration of stay in vicinity to the single white stripe and the blue vertical stripes at day 4 in two consecutive recordings (Fig. R14a, b).**

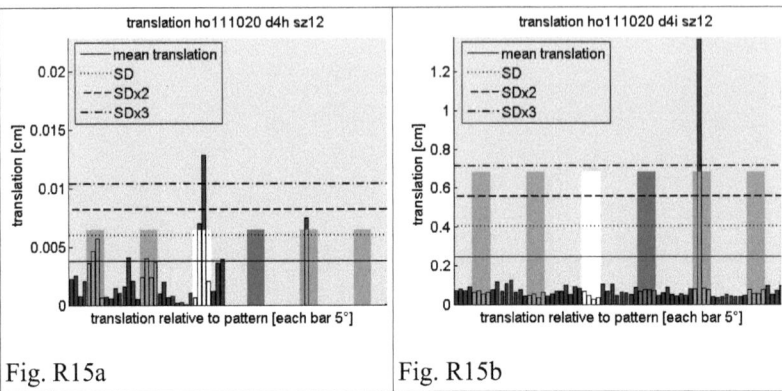

Fig. R15a   Fig. R15b

Fig. R15: Forward running (translation) for hornet 111020 at the fourth day in the virtual environment. The white and dark grey bar indicate the position of the white and black vertically presented stripes in the virtual szene 12. The light grey bars indicate blue stripes. Each stripe has a width of 20°, the x-axis represents the 360° virtual panorama. The slim bars indicate the animals translation in the virtual space (each bar 5°). The horizontal lines indicate mean translation and mean plus 1, 2 and 3 times standard deviation (SD). **Please note the increased amount of translation in vicinity to the single white stripe and the blue vertical stripes in fig. R15a in comparison to the black stripe. In the following test, the animal is preferentially walking towards a blue stripe. The animal enlarges the selected cue by means of translatory walking activity on the ball.**

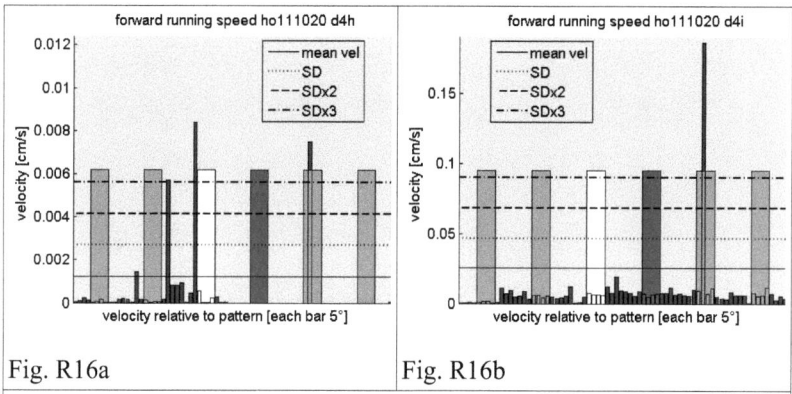

Fig. R16a Fig. R16b

Fig. R16: Forward running (translation) velocity for hornet 111020 at the fourth day in the virtual environment. The white and dark grey bar indicate the position of the white and black vertically presented stripes in the virtual szene 12. The light grey bars indicate blue stripes. Each stripe has a width of 20°, the x-axis represents the 360° virtual panorama. The slim bars indicate the animals translation in the virtual space (each bar 5°). The horizontal lines indicate mean translation and mean plus 1,2 and 3 times standard deviation (SD).**The forward running speed is maximal towards the white stripe and the blue stripes (Fig. R16a, b) in comparison to the translation activity towards the black stripe in both recordings at the same experimental day.**

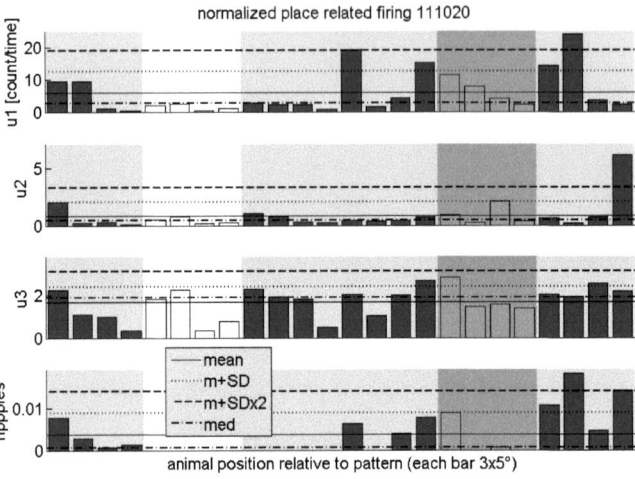

Fig. R17: Place related firing in units 1 and 2. The white and grey bar indicate the position of the white and black vertically oriented stripes in the szene Nr. 7, which was presented here. It consisted of 6 alternating black and white stripes, each with a widh of 20°. The slim bars represent the normalized firing rate of the animal in 5° of the virtual space, respectively. Since the space is normalized to one black and one white stripe (120° of the whole VE), each bar represents 5°x3=15° of the panorama. The horizontal lines indicate mean firing rate and mean rate plus1 and 2 times standard deviation (SD) as well as the median (med). The firing rate is normalized to the duration of stay in each spatial segment. The same was done with the occurence of ripple events, shown in the bottom row. Units 1 and 2 show place related firing, whereas unit 3 (third row) does not.

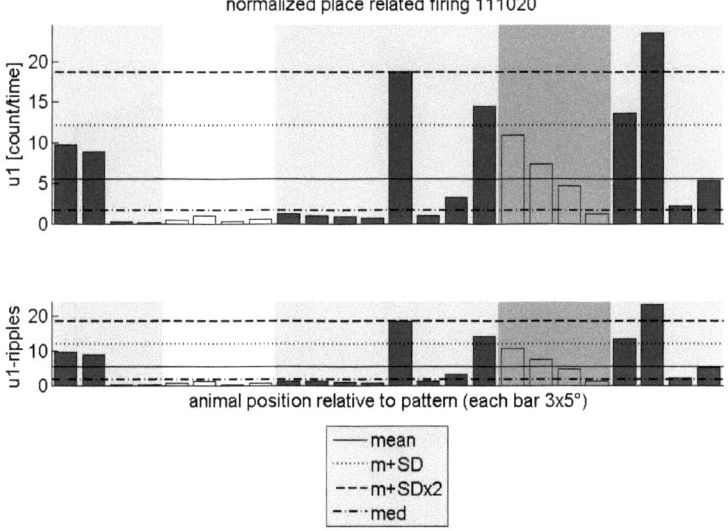

Fig. R18: Correction for ripple events. The white and the grey bar indicate the position of the white and black vertically oriented stripes in the szene Nr. 7, which was presented here. It consisted of 6 alternating black and white stripes, each with a widh of 20°. The slim bars represent the normalized and corrected firing rate for the animal in 5° of the virtual space, respectively. Since the presented space is normalized to one black and one white stripe (120° of the whole VE), each bar represents 5°x3=15° of the panorama. The horizontal lines indicate mean firing rate and mean rate plus1 and 2 times standard deviation (SD) as well as the median (med). The firing rate is normalized to the duration of stay in each spatial segment (top row). Additionally, the mean influence of ripple events on the firing rate was substracted, if ripples occured in a given 15°window of the panorama (bottom row).

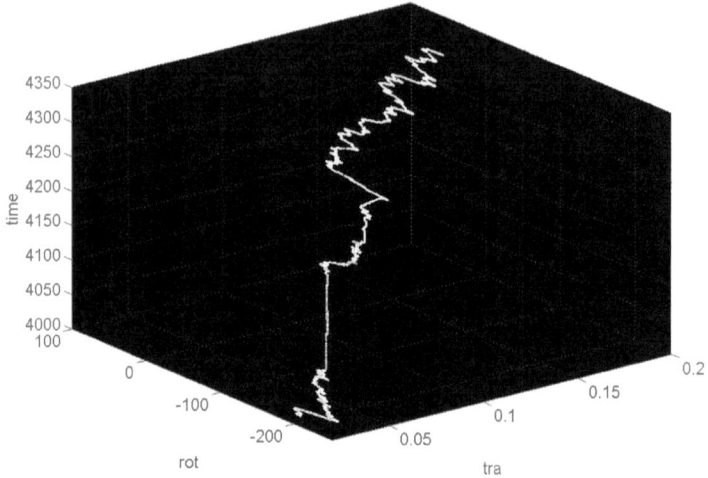

Fig.R19: 3D plot of virtual trajectory and place related spike pattern at day 2c. The translation is plottet on the axis „tra" [mm], rotation on the axis „rot" [°] and time [s] on the z-axis. The activity of unit 1 is colour coded. Cyan means activity, white means no spiking activity. Please note the higher amount of unit 1 activity during a pause at the beginning of the trajectory. Zig-zag movements are visible during activity phases. A black stripe of the virtual szenario is in the range of -140 to -160°, the next white stripe starts at -200°. The hornet was navigating in szenario7 with alternating black and white stripes as described in the methods. The stripe width was 20°, respectively and the interspace 40°.

Results for the freely running honeybee. A forager honeybee with a chronically implanted micro plug was recorded for two consecutive days and one night. The bee was not able to fly but had full walking capacity. After recovery from the implantation procedure, the bee was released in a 10x10 cm transparent walking arena, with bees wax on the ground. The arena had 4 functional quadrants (Fig. R20, also used for analysis). Quadrant 1 (Q1) yielded a sugar solution on top of a blue piece of paper and Quadrant 2 (Q2) a small dish with water on top of a yellow piece of paper. Quadrant 3 (Q3) was empty and in the edge of Quadrant 4 (Q4), a small tube with beeboost (Queen pheromone) was placed. A representative video of approximately one hour duration was anlyzed with respect to possible correlations between behavior and electrophysiology as well as spatial preferences in the arena. During this first hour of exploration, the bee spent 63% of the time in Quadrant 3, 25 % in Q4 and 11% in Q1. Q1 was also the site of release. During the same time, 22 ripple cycles occured in Q3, 8 in Q4 and 4 in Q1. Normalized to the amount of time, spent in these areas, there have been 36 ripple cycles in Q1, 35 in Q3 and 32 in Q4. The ripples occured most often during pauses in navigation, or moderate behavioral actions, like turning on the point or moving the abdomen or hind legs. The ripples are also not associated with a special head direction. The pattern of the ripples is similar to that, observed in the hornet (Fig. R21). During this 1h recording at the first day, the ripple cycles had a duration from 0.1s to 11s. Again, the ripples are characterized by a more or less steep onset (0.01s until the peak potential is reached). The whole duration of one ripple (positive or negative deflection of one cycle) was approximately 0.05s. The duration of ripple cycles is not statistically different between sleep phases and activity. There is a tendency towards a shorter duration of ripple cycles during sleep phases (mean sleep ripple cycle duration at d1f: 1.84s, mean awake ripple cycle duration at d1f: 3.65s). Additionally there is also a trend for a higher amount of ripples in Quadrant 4 where the ripples initially occured, but only for the non-sleep phases. The ripples were normalized to the amount of time, the bee spent in a single Quadrant. For d1f (night recording from 10:50- 11:30 p.m.) 55 normalized ripple cycles occured in Q2, 68 during non-sleep phases in Q4 and 54 during sleep in Q4. This effect is not seen in an earlier recording. At d1c (8:50-9:40 p.m.) the normalized ripple cycles per Quadrant were: Q1:36, Q3:35 and Q4:32.

During sleep phases the increase in synchronous spike frequency of unit 2 (fig. R22) is greater than during wakeful states, like pauses during navigation (fig. R23).

Quadrant 1 (Q1), blue, sugar water	Quadrant 2 (Q2), yellow, water
Quadrant 3 (Q3), empty	Quadrant 4 (Q4), Bee boost

Fig. R 20: For analysis the walking arena (10x10cm) was divided into 4 quadrants (Q), with different sensory stimuli.

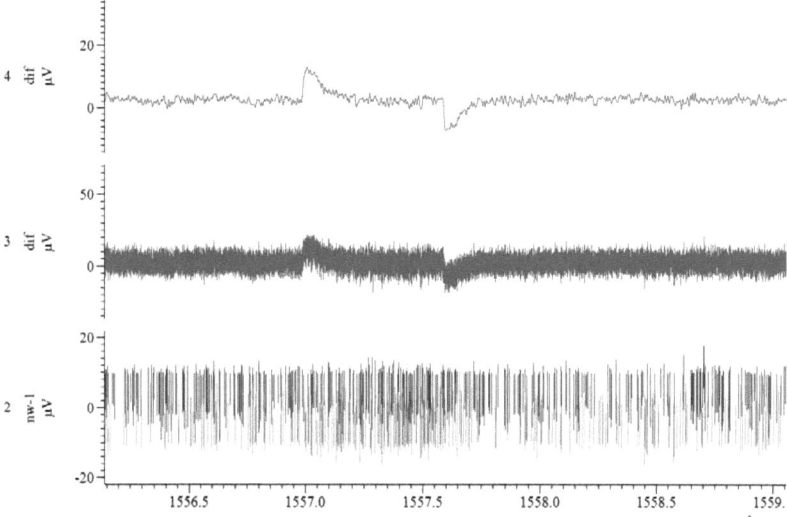

Fig. R21: Ripple cycle during navigation. The second row (Cannel 3) is the original recording with one ripple cycle (onset at approximately second 1557, approximately 9:00 p.m.). To extract the ripple events, the channel was low pass filtered (top row). The bottom row indicates an increase in simultaneously recorded unit activity (unit1 in blue and unit 2 in green).

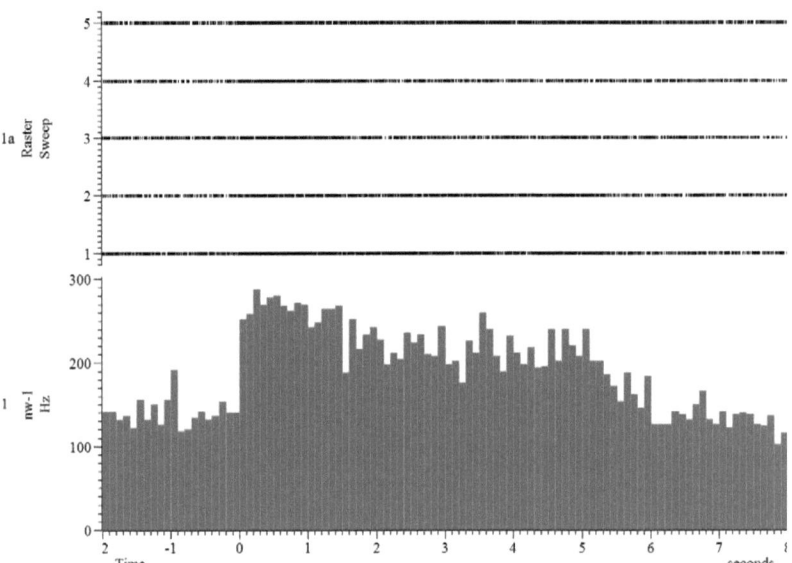

Fig. R22: PSTH for ripple cycles during sleep. 5 consecutive ripple cycles during a sleep phase. The cycles are aligned at the the positive deflection of the baseline potential (ripple on-phase, +) at second 0 in the PSTH. The spikes are binned for 0.1s. Since the ripples have different durations, the decline of frequency is rtificially not as steep as the onset. The raster plot on top displays spike events as lines. During sleep, the frequency of unit 2 increases from approximately 110 Hz to nearly 300Hz peak frequency during the ripple cycles.

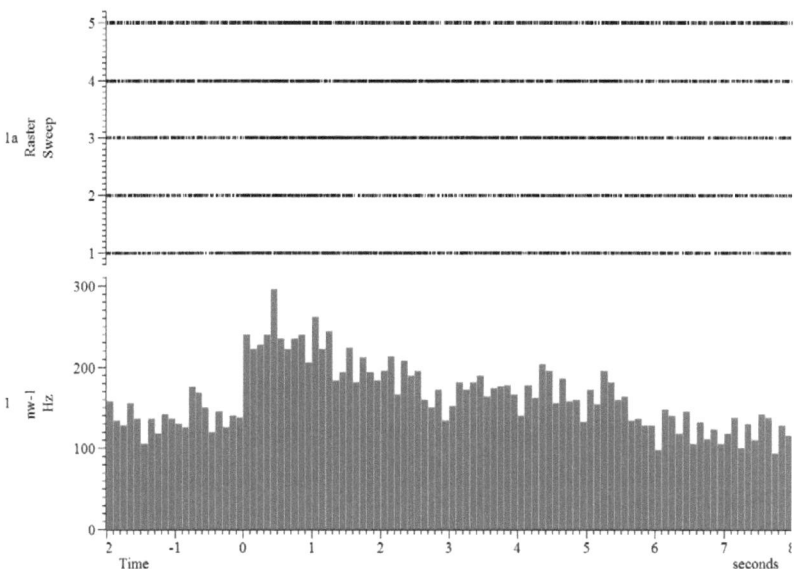

Fig. R23: PSTH for ripple cycles during non-sleep states. Five consecutive ripple events during locomotion or, more often, pauses during navigation. The ripples are aligned with the positive baseline deflection (ripple cycle on-phase, +) at second 0 in the PSTH. The spikes are binned for 0.1s. Since the ripples have different durations, the decline of frequency is not as steep as the onset. The raster plot on top displays spike events as lines. **During sleep, the frequency of unit 2 increases from approximately 110Hz to nearly 300Hz peak frequency during the ripple cycles.**

Discussion. The data presented here, give a first evidence for ripple-like potentials during virtual and real navigation in insects. The ripples, recorded in a virtually navigating hornet, are functionally related to the ripples described in the rodent brain as they are associated with high unit activity in place related firing patterns (fig.R19). The ripple-like events in insects share some prominent features with the hippocampal ripples (fig.R1). For example, the ripple cycles are correlated with an increase in unit activity. The units which are firing in synchrony with the ripples show a place specific activation pattern in the virtual environment (fig.R17). The ripples are occuring in the whole virtual and real (arena) space but also more likely in the place fields of the associated units. One should keep in mind, that a place field in flying insects might be larger than the arena for the freely walking honeybee, used here. This is not the case in the virtual environment, where occlusion and motion parallax simulate a rather big distance for the virtually navigating hornet. Interestingly, the ripples occur most often in pauses during navigation and sleep states. This is a central feature of hippocampal ripples. The mushroom body ripples do not only raise the question for a basic function in the insect brain and functional relations to the mammalian brain. They are also another strong hint for the occurence of cognitive maps (as defined by Tolman, 1948) in insects. In rodents, place cell activity is interpreted as an integral part of such a mind map (O'Keefe and Dostrowski, 1971; Wilson and McNaughton, 1993). Since the ripples are associated with replay and in this context consolidation of spatial memory loads, they are associated with mnemonic processes which are necessary for the storage of cognitive maps (Girardeau et al., 2009). The occurence of cognitive maps is controversely discussed for insects. On the one hand it was mentioned that cognitive maps are simply not necessary for insect navigation, because the experimental results could be fully described by the existence of sun compass based orientation (Wehner&Menzel, 1990, Cruse&Wehner, 2011). On the other hand it has been shown in experiments with free-flying honeybees that map-like structures are likely to explain the direct route from an unfamiliar, but relatively near release site to the trained feeder rather than back to the hive (Gould, 1986). Menzel et al. (2005) measured whole flights with a harmonic radar. The complexity of the observed flight patterns led them to the suggestion, that different navigation strategies are combined, including spatial knowledge based on cognitive maps. Obviously it is not easy to proof basal brain mechanisms on the basis of mere behavioral data. Therefore, the electrophysiological evidence of sharp waves with associated „high" phases in some units, which show also place related firing in a virtual reality is maybe the missing link to a long debated issue. The data presented here, give strong evidence for the occurence of cognitive maps in the insect brain.

It remains the question if replay phenomena are restricted to spatial memory loads or if they are a more general phenomenon to produce long lasting duplicates of otherwise transient sensory inputs. Replay might generally be an important step towards cross modal computations, filtering and feature extraction, which are surely necessary to adapt behavioral outputs to environmental necessities. It is obvious that the experimental reduction to single sensory modalities is an artificial situation with rare occurence in real environments. If a honeybee is searching for nectar, it uses at least visual, olfactory and sensory cues. Jacobs et al. (2011) suggest an olfactory spatial hypothesis (or its revival). The olfactory neuropile in vertebrates as well as in invertebrates is primarily dedicated to navigational needs and not odor discrimination, which is a prerequisite for the latter. Indeed, the interplay between sensory modalities has already been observed in experiments with visual conditioned stimuli (CS)in classical conditioning in restrained honeybees. It has been shown that the PER towards an visual stimulus is only elicited when the antennae have been cut (Kuwabara, 1957). New reslults indicate, that this effect is dependent on age and younger honeybees show good learning results in visual classical conditioning with intact antennae (Dobrin&Fahrbach, 2012). The suppression of the proboscis extension reflex (PER) to color stimuli in forager bees might be an adaptation to the needs of navigation – especially since it is clearly documented, that forager bees are able to discriminate these colours and select them according to an associated sugar reward (von Frisch, 1914). Interestingly, the mushroom body ripples have never been observed in experiments without translatory feedback. As is also true for the research in rodents, the integration of a wide field of sensory modalities during the experimental procedure requieres not only more complex analysis techniques but also a growing number of recording channels. To unravel the function of mushroom body ripples and neural correlates of complex behavioral actions like decision making it will be necessary to perform recordings with much more electrodes than used here. A true mapping is required and therefore microdrives as they are used in rodents (Kloosterman et al., 2009). The insect brain is simply to small to perform this kind of recording with metal wires. One option for the future can be carbon wires as they are already used for extracellular recordings (Piironen et al., 2011).

As already described, the mushroom body ripples share some features with the hippocampal ripples. This is also interesting from an evolutionary point of view. It seems likely that the occurence of ripple-like brain states precedes the occurence of the hippocampus. Interstingly, the mushroom body is often discussed as „hippocampus-like" structure in invertebrates (Mizunami et al., 1998). It is known that basic physiological properties have striking similarities across taxa (Kandel& Abel, 1995, Menzel&Manz, 2005, Hammer, 1993). This has not been shown for replay phenomena or place related firing, yet. Nevertheless, oscillating neuronal activity has often been described in the mushroom body, as well as reverberating patterns (Gronenberg, 1987).

The parsimonious explanation of ripple-like events and place related firing patterns is a cognitive map in insects or at least hornets and honeybees.

Interestingly, the mushroom body developmental gene expression patterns are most lokely momologous to gene expression patterns in the mammalian hippocampus (Tomer et al., 2010). These data, based on a newly developed technique, strongly indicate that the mushroom body and the hippocampus have a common link, before deuterostomes and protostomes divorced to different developmental lineages, about 600 million years ago (Tomer et al., 2010). Therefore, ripples/sharp wave-like structures and replay might be evolutionary far older than the hippocampus, where they have been described first (Buzáki et al., 1992). This is in line with the assumption, that ripples and replay are fundamental processes during memory consolidation (Girardeau et al., 2009) with underlying molecular circuits which are already found in molluscs (Kandel&Abel, 1995).

References:

Buzsáki, G., Horváth, Z., Urioste, R., Hetke, J., and Wise, K. (1992). High frequency network oscillation in the hippocampus. Science **256**: 1025–1027.

Cruse, H., Wehner, R. (2011) No need for a cognitive map: decentralized memory for insect navigation. PloS Comput. Biol. **7(3):** e1002009.

Denker, M., Finke, R., Schaupp, F., Grün, S., Menzel, R. (2010) Neural correlates of odor learning in the honeybee antennal lobe. Eur. J. Neurosci. **31:** 119-33.

Dobrin SE, Fahrbach SE (2012) Visual Associative Learning in Restrained Honey Bees with Intact Antennae. PLoS ONE **7(6):** e37666. Doi:10.1371/journal.pone.0037666.

Dombeck, D.A., Harvey, C.D., Tian, L., Looger, L.L., Tank, D.W. (2010) Functional imaging of hippocampal place cells at cellular resolution during virtual navigation. Nat. Neurosci. **13:** 1433-1440.

Gould, J.L. (1986) The Locale Map of Honey Bees: Do Insects Have Cognitive Maps? Science **232:**
861-863.

Gronenberg, W. (1987) Anatomical and physiological properties of feedback neurons of the mushroom bodies in the bee brain. Exp. Biol. **46:**

115-125.

Erber, J., T. Masuhr, and R. Menzel. (1980) Localization of short-term memory in the brain of the bee Apis mellifera. Physiol. Entomol. **5:** 343–358.

Ferguson, J.E., Boldt, C., Redish, A.D. (2009) Creating low-impedance tetrodes by electroplating with additives. Sens. Actuators. A Phys. **156:** 388-393.

Friedrich, M., Tautz, D. (1995) Ribosomal DNA phylogeny of the major extant arthropod classes and the evolution of myriapods. Nature **376:** 165-167.

von Frisch, K. (1914) Der Farbensinn und Formensinn der Bienen. Zool. Jb., Abt. Allg. Zool. u. Physiol. **35:** 1-238.

Gillner, S., Mallot, H.A. (1998) Navigation and Acquisition of Spatial Knowledge in a Virtual Maze. J. Cogn. Neurosci. **10:** 445-463.

Girardeau, G., Benchenane, K., Wiener, S.I., Buzsáki, G., Zugaro, M. B. (2009) Selective suppression of hippocampal ripples impairs spatial memory. Nat. Neurosci., **12:** 1222-1223.

Harvey, C.D., Collman, F., Dombeck, D.A., Tank, D.W. (2009) Intracellular dynamics of hippocampal place cells during virtual navigation. Nature **461:** 941-946.

von Heimendahl, M., Rao, R.P., Brecht, M. (2012) Weak and Nondiscriminative Responses to Conspecifics in the Rat Hippocampus. J. Neurosci. **32:** 2129-2141.

Heisenberg, M., A. Borst, S. Wagner, and D. Byers. (1985) Drosophila mushroom body mutants are deficient in olfactory learning. J. Neurogenet. **2:** 1–30.

Höllscher, C., Schnee, A., Dahmen, H., Setia, L., Mallot, H.A. (2005) Rats are able to navigate in virtual environments. J. Exp. Biol. **208:** 561-569.

Hussaini, S.A., Kempadoo, K.A., Thuault, S.J., Siegelbaum, S.A., Kandel, E.R. (2011) Increased Size and Stability of CA1 and CA3 Place Fields in HCN1 Knockout Mice. Neuron **72:** 643-653.

Kandel, E. and T. Abel (1995) Neuropeptides, adenylyl cyclase, and memory storage. Science **268:** 8225–8226.

Kenyon, F.C. (1896) The meaning and structure of the so-called "mushroom bodies" of the hexapod brain. Am. Nat. **30:** 643–650.

Kloosterman, F., Davidson, T.J., Gomperts, S.N., Layton, S.P., Hale, G., Nguyen, D.P., Wilson, M.A. (2009) Micro-drive Array for Chronic in *vivo* Recording: Drive Fabrication. JoVE.26. Http://jove.com/index/Details.stp? ID=1094, doi: 10.3791/1094

Kuwabara, M. (1957). Bildung des bedingten Reflexes von Pavlovs Typus bei der
Honigbiene, Apis mellifica. J. Fac. Sci. Hokkaido Univ. Ser. VI Zool. **13**: 458-464.

Lindauer, M. (1959) Angeborene und Erlernte Komponenten in der Sonnenorientierung der Bienen. Z. Vergl. Physiol. **42**: 43-62.

Liu, L., Wolf, R., Ernst, R., Heisenberg, M. (1999) Context generalization in *Drosophila* visual learning requires the mushroom bodies. Nature, **400**: 753–756

Martin, J.-R., Ernst, R. &Heisenberg, M. (1998) Mushroom bodies suppress locomotor activity in Drosophila melanogaster. Learn. Mem. **5**: 179-191.

Masuhr, T. (1976) Lokalisation und Funktion des Kurzzeitgedachtnisses der Honigbiene (Apis mellifera L.). Ph.D. Dissertation. Fachbereich Biologie, TH Darmstadt, Germany.

Menzel, R. (1975) Electrophysiological Evidence for Different Colour Receptors in One Ommatidium of the Bee Eye. Z. Naturforsch. **30 c**: 692-694.

Menzel, R, Manz, G. (2005) Neural plasticity of mushroom body-extrinsic neurons in the honeybee brain. J.Exp.Biol. 208: 4317-4332.

Menzel, R., et al. (2005) Honey bees navigate according to a map-like spatial memory. PNAS **102**: 3040-3045.

Menzel, R., Giurfa, M. (2006) Dimensions of Cognition in an Insect, the Honeybee. Behav. Cogn. Neurosci. **5**: 24-40.

Mizunami, M., J.M. Weibrecht, and N.J. Strausfeld. (1993) A new role for the insect mushroom bodies: Place memory and motor control. In Biological neural networks in invertebrate neuroethology and robotics (ed. R.D. Beer, R.E. Ritzman, and T. McKenna), pp. 199–225. Academic Press, New York, NY.

Mizunami, M., R. Okada, Y.-S. Li, and N.J. Strausfeld (1998) Cockroach mushroom bodies: Activity and identity of neurons in freely moving animals. J. Comp. Neurol. **402**: 501–519.

Okada, R., Rybak, J., Manz, G., Menzel, R. (2007) Learning-Related Plasticity in PE1 and Other Mushroom Body-Extrinsic Neurons in the Honeybee Brain. J. Neurosci. **27**, 11736-11747.

O'Keefe, J., and Dostrovsky, J. (1971). The hippocampus as a spatial map. Preliminary evidence from unit activity in the freely-moving rat. Brain Res. **34**: 171–175.

Peng, Y., Xi, W., Zhang, W., Zhang, K., Guo, A. (2007) Experience Improves Feature Extraction in *Drosophila*. J. Neurosci. **27**: 5139-5145.

Piironen, A., Weckström, M., Vähäsöyrinki, M. (2011) Ultrasmall and customizable multichannel electrodes for extracellular recordings. J. Neurophysiol. **105:** 1416–1421.

Regier, J.C., Shultz, J.W., Kambic, R.E. (2005) Pancrustacean phylogeny: hexapods are terrestrial crustaceans and maxillopods are not monophyletic. Proc.R.Soc. **272:** 395-401.

Rybak, J., Menzel, R. (1998) Integrative Properties of the Pe1-Neuron, a Unique Mushroom Body Output Neuron. Learn. Mem. **5:** 133-145.

Strube-Bloss, M.F., Nawrot, M.P., Menzel, R. (2011) Mushroom Body Output Neurons Encode Odor-Reward Associations. J. Neurosci. **31:** 3129-3140.

Taki, M., Iyoshi, S. Ojida, A., Hamachi, I., Yamamoto, Y. (2010) Development of Highly Sensitive Fluorescent Probes for Detection of Intracellular Copper(I) in Living Systems. J. Am. Chem. Soc. **132:** 5938-5939.

Tolman, E. C. (1948) Cognitive maps in rats and men. Psychol.Rev. **55:** 189-208.

Tomer, R., Denes, A.S., Tessmar-Raible, K., Arendt, D. (2010) Profiling by Image Registration
Reveals Common Origin of Annelid Mushroom Bodies and Vertebrate Pallium. Cell **142:** 800-809

Van Swinderen, B., Greenspan, R.J. (2003) Salience modulates 20-30 Hz brain activity in *Drosophila*. Nat. Neurosci. **6:** 579-586.

Wehner, R. & Menzel, R. (1990) Do insects have cognitive maps? A. Rev. Neurosci. **13:** 403-414.

Wilson, M.A., and McNaughton, B.L. (1993). Dynamics of the hippocampal ensemble code for space. Science **261:** 1055–1058.

Wolf, R (1998) *Drosophila* Mushroom Bodies are Dispensable for Visual, Tactile and Motor Learning. Learn. & Mem. **5:** 166-178.

Wolf, R.,Heisenberg, M. (1991) Basic organization of operant behavior as revealed in *Drosophila* flight orientation. J. Comp. Physiol. A **169:** 699-705.

Zhang, K., Guo, J., Peng, Y., Xi, W., & Guo, A. (2007). Dopamine-mushroom body circuit regulates saliency-based decision-making in Drosophila. Science, **316:** 1901–1904.

6. Drug effects on walking activity in the virtual environment

Introduction. Do Cocaine, Octopamine or Ritalin (Methylphenidate) alter the walking activity in honeybees, walking stationary on a treadmill? Is the neural activity changed by these drugs? Cocaine, a plant alcaloid, is known as reuptake inhibitor of dopamine in the vertebrate as well as the insect (Ritz et al., 1987, Corey et al., 1994) nervous system. In contrast to vertebrates, cocaine inhibits also the octopamine reuptake in invertebrates (Gallant et al., 2003). Furthermore, dopamine and octopamine transporters are comparably insensitive to cocaine. In contrast to the human dopamine transporter sensitivity, one order of magnitude more cocaine is needed to achieve the same effect of dopamine reuptake inhibition in invertebrates (Gallant et al., 2003). Biogenic amines are a group of mono- and diamines which are synthesized by decarboxylation of aminoacids. They are built in neurosecretory cells and have a widely distributed function as hormones, neuromudulators or neurotransmitters (Römpp, Lexikon der Chemie). It has been suggested, that octopamine is a functionally homologue to noradrenaline in vertebrates (Roeder, 1999). This suggestion arised not only due to structural similarities but also from similar physiological effects in their respective role as neurotransmitter/neuromodulator for example in the stress response (Roeder, 1999). Octopamine is synthesized by beta-hydroxylation of tyramine (Monastirioti, 1996), which also acts as neurotransmitter/neuromodulator (Lange, 2009). Tyramine is produced by decarboxylation of tyrosine (Livingstone&Tempel, 1983). Theoretically it exists also the possibility that tyramine is synthesized of dopamine by dehydroxylase activity, but an in vivo function has not yet been shown (Walker&Kerkut, 1978). Recently it as been shown that one octopamine receptor (AmOA-1) is expressed in the calyces, the lobes, central body and especially in optic neuropiles of the bee brain (Sinakevitch et al., 2011). Furthermore, these receptors are found in inhibitory feedback neurons of the PCT and therefore some colocalizations with anti GABA stainings exists (Sinakevitch et al., 2011). The AmOA-1 receptor belongs to the class of octopamine receptors, which lead to changes in the intracellular calcium concentrations, whereas the activation of the other class of octopamine receptors lead to the activation of adenylyl cyclases and an increase in intracellular cAMP concentrations (Blenau&Baumann, 2001). In the peripheral nervous system of insects, octopamine increases the energy metabolism in flight muscles and other peripheral organs and is generally associated with stress (fight or flight) responses (Goosey&Candy, 1980, Verlinden et al., 2010). In the locust visual system octopamine modulates arousal by dishabituation (Stern, 1999). Martin Hammer published in 1993 his finding of octopaminergic reward neurons in the honeybee brain. He was also able to demonstrate, that the unconditioned stimulus during classical conditioning of honeybees with the proboscis extension reflex (PER) could be replaced by octopamine injection into the antennal lobes and the mushroom bodies of the brain, to achieve olfactory learning. Octopamine and tyramine increase the bees responsiveness to sugar, whereas dopamine reduces the sugar responsiveness (Scheiner et al., 2002). Octopamine and tyramine alter locomotor activity in honeybees (Fussnecker et al., 2006). Freely flying honeybees, fed with octopamine, solved in sucrose or treated with octopamine, solved in DMF, on the thorax, show an increased dance duration rather than dance likelihood. This finding was also true for pollen collecting forager bees, thus excluding a solely sugar related effect (Barron et al., 2006). Furthermore octopamine has also influences on the social behavior, for example the division of labor and is thought to play a role in the switch from indoor work to foraging tasks in honeybees (Schulz et al., 2002). Octopamine plays a crucial role in appetitive learning (Hammer, 1993) and reduces aversive learning in a place-preference paradigm (Agarwal et al., 2011). Dopamine seems to influence appetitive as well as aversive learning (Kim et al., 2007). New findings show a differential expression of octopamine and dopamine receptors in different populations of mushroom body intrinsic neurons during the lifetime of a honeybee, which suggests an additional level of complexity for the mechanisms, which are regulated by biogenic amines (McQuillan et al., 2012). Three different dopamine receptors are expressed in the mushroom body of honeybees. *Amdop*1 (Blenau et al., 1998) and *Amdop*2 (Humphries et al., 2003) express G-protein coupled receptors, which may increase intracellular levels of cAMP. AmDOP1 acts additionally via increases in intracellular Ca concentrations (Grohmann et al., 2003). The activation of *Amdop*3 receptors reduces intracellular cAMP concentrations (Beggs et al., 2005) or sometimes leads to an increase in cAMP depending on the cellular milieu (Clark&Baro, 2007). As described before, cocaine affects octopamine as well as dopamine reuptake in invertebrates (Gallant et al., 2003). It was shown that cocaine increases the likelihood and the rate of dance behavior in honeybees, but does not affect the general locomotor activity (Barron et al., 2009). Barron et al. (2009) were also able to demonstrate an increase in sucrose responsiveness after cocaine treatment as well as negative effects on learning after withdrawal of cocaine in previously chronic treated bees. Ritaline is used to treat attention-deficit/hyperactivity disorder (ADHD) in human, especially children (Accardo&Blondis, 2001, Frölich et al., 2012). Methylphenidate is a cyclized derivative of amphetamine (Teo et al., 2002) and acts as reuptake inhibitor of dopamine, very similar to cocaine but with a longer efficacy in the human brain for methylphenidate (Krause et al., 2000, Volkow et al., 1999). In a recent study in rats it was shown that methylphenidate has complex influences to short and long term changes in the firing rate of neurons in the nucleus accumbens (Chong et al., 2012). Methylphenidate rescues attention deficits in *Drosophila* memory mutants (Van Swinderen&Brembs, 2010). As has been shown, numerous studies on the pharmacology and genetics on biogenic amines and drugs are available. In contrast, electropysiological studies for long term changes during drug adminstration are rare. This study aims for bridging this gap of knowledge via an insect model. One single identified neuron in the ventral part of the mushroom body of the honeybee, the PE1 neuron, is known for learning related placticity and characteristic burst phases (Mauelshagen, 1993). Therefore, the extracellular electrodes were placed in the region of the PE1 neuron at the ventral border of the mushroom body alpha lobe.

Material and Methods.

Virtual Environment Setup. Spherical treadmill, geometry of the virtual environment and overall setup were described by Höllscher et al. (2005), apart from the following changes (see also Fig. M1): The treadmill consisted of a Styrofoam ball (10 cm diameter) placed in a half-spherical plastic cup with several, symmetric located holes through which a laminar air flow passed and let the ball float on air. Laminarity of the air flow was supported by a long (12 m) tube from the well regulated fan to the ball. Low static electrification of the ball and humidity for the animal was achieved by blowing the air into a box with water. Buoys prohibited corrugation inside the box.

The beamer Epson EMP-TW 700 (digital scanning frequency: pixel clock: 13,5-81 MHz, horizontal sweep: 15-60 kHz, vertical sweep: 50-85 Hz) was positioned above the Faraday cage and illuminated the inner surface of a cone shaped screen (height 60 cm, bottom diameter 7 cm, top diameter 75 cm) via a large surface mirror and a Perspex window (Fig.1). A Fourier power analysis of the beamer projection via a photodiode (thanks to Uwe Greggers) revealed a main part 210 Hz frame rate. Less than 10% 100Hz and 60Hz have been measured, respectively (measurement 15.2.2010, kindly provided by Uwe Greggers). The inner surface of the cone consisted of white paper. Patterns projected onto this screen were distorted such that they appeared undistorted when seen by the bee (BeeWorld, programmed by Sören Hantke).

During an experiment, the Faraday cage was closed. A webcam (Logitech, Morges Gesellschaft) positioned above imaged the head of the animal via a 500 mm mirror objective allowing observation of the animal during the experiment.

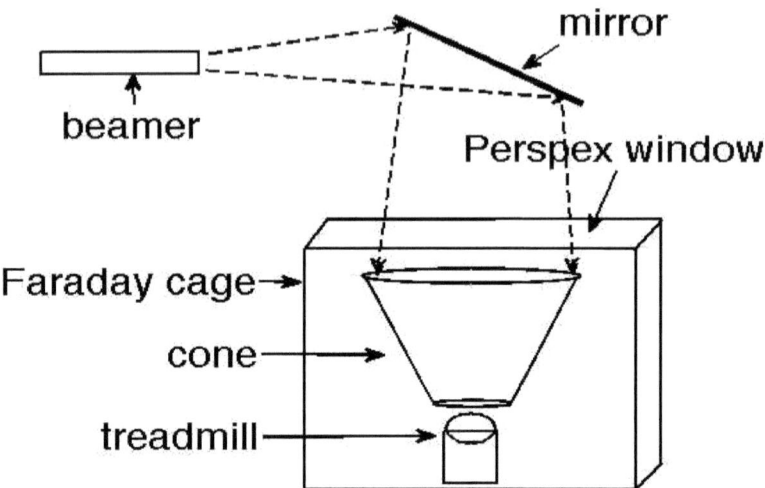

Fig. 1: The light was projected from a beamer onto a mirror and from the mirror onto the inner surface of a cone shaped sreen. The screen was placed inside a Faraday cage with a Perspex window on top, to allow for a passage of the light. The screen could be lifted to place the animal on the styrofoam ball (treadmill).

Control of the virtual environment and experimental procedure. The virtual environment and the recognition of the hornets movement was under the control of the customs written program BeeWorld (Sören Hantke). It was implemented in Java by using OpenGL-Bindings for Java (LWJGL). The movement of the ball, initiated by the walking bee, was detected by two optical computer mice as they are used for computer games (Imperator, Razer Europe GmbH; G500, Logitech Europe S.A.). The mice were accurately positioned under 90° at the equator of the styrofoam ball and precisely aligned with x/y micro drives. The animal was able to control the virtual scenery by rotatory movements of the ball. Translatory movements magnified the objects on the screen as if the animal would approach. Multiple scenarios were programmed. They were realized as xml-files containing the positions, width and color (RGB) of a variable number of vertically oriented stripes. These stripes were positioned around the bee. Scenario7 consisted of alternating black and white, vertically oriented stripes with an angular width of 20° and an grey interspace of 40°. In the figures presented here, the redundant pattern is standardized on one black and one white pattern in an angle of 120°. Scenario12 consisted of one black, one white and four blue stripes with the same dimensions as in scenario7.

The field of view of a camera in OpenGL is limited to 179°, the scenarios projected onto the screen, however, simulated a 360° world. In the Bee-World program a four texture renderer (texturerenderer) was used to create a 360° camera. Data of walking traces were synchronized with the data from spike recordings which were collected with an analog to digital converter (micro3, CED, Cambridge Electronic Design, 30 kHz sampling frequency per channel). A Silicon NPN Phototransistor (BPY 62) directed at the screen detected a short light signal under the control of the BeeWorld program and fed it into the ADC input of the analog to digital converter. The pulse was recorded in a spike2 channel and allowed for precise timing between the spike and walking data, recorded with different computers.

A skyline in the background of the scenario consisted of vertically oriented stripes of different hues of grey and different length. A checker board pattern was projected on the plain floor around the hornet. The angular rotation of this checker board pattern was equal to the angular movement of the ball simulating a respective movement of the floor directly below and around the animal. The angular rotation velocity of the stripe pattern was set to 75% of the checker board pattern, and a skyline pattern projected onto the screen together with the stripe pattern moved with 50% speed of the checkerboard pattern. Thus these three patterns simulated depth information by creating a signal of different motion parallax as seen by the stationary walking insect.

Electrophysiology in stationary walking *Hymenoptera*. Four insulated copper wires (Elektrisola) with a diameter of 0.015mm each were coiled and fixed with superglue or heated (210° for 3 minutes have been appropriate to fuse the polyurethane insulation of the wires without producing an electric shortage). These techniques increased the stability of the coiled electrode bundles and facilitated tissue penetration.

The best signal to noise ratio was achieved with Teflon insulated silver ground wires with a diameter of 0,125 mm (WPI, Berlin, D) implanted into the abdomen or Teflon insulated Platinum/Iridium wires of 0.025mm diameter (advent, Enysham, Oxford, UK) which were thin enough to be implanted in the brain, compound eye or ocelli. The insulation at the tip of these wires was removed mechanically with a fine forceps.

The tip of the tetrode was cut with an fine scissors. The single ends of the copper wires were dipped into hot solder to remove the insulation and then connected with silver conductive paint (electrolube) to the female side of the pins of an IC socket. After drying the coiled ends of the electrodes were electroplated as described in the low-impedance plating procedure with gold and PEG (Ferguson et al., 2009) using the electroplating device NanoZ (Neuralynx).

Preparation of the animals. The Bees were caught at the hive entrance. They were chilled on ice and fixed temporarily in a small tube with modelling clay. A small piece of plastic tube or rubber foam was fixed with dental wax on the thorax as holder for the stationary running animal on the treadmill. A window between the compound eyes, antennae and ocelli was cut into the head capsule. The tip of the electrode was fixed to a fine forceps which was mounted on an external micromanipulator. The electrode was inserted into the brain, while the animal was still harnessed in the tube. After placing the electrodes in the selected brain area (ventral aspect of the alpha lobe of the mushroom body) under visual control, the electrode was fixed with two component silicone elastomer (kwik sil, WPI) onto the brain and the head capsule. After hardening of the kwik sil, the electrode was released from the external micromanipulator and additionally fixed inside a small slit in the plastic tube/rubber foam holder on the thorax by forming a small loop from the head backwards to the thorax. About 5 minutes later, the bee was released from the tube by grabbing the plastic tube/rubber foam mounted onto the thorax with a forceps and pushing the head with another forceps slowly backwards to facilitate the animals release from the tube. The rubber foam had a slit and the electrodes could be fixed without silicone elastomer. Fixation of the electrodes, however, was crucial for stable and long lasting recordings. Afterwards, the bee was clipped by the plastic tube or piece of rubber foam on its thorax to an alligator clip, attached to the electrode holder. Longer electrodes were additionally fixed with modelling clay to the electrode holder to prevent reachability of the fine wires from the range of the bee's legs. Especially during the transfer to the setup after the release from the fixation tube and during the first minutes on the ball the animals often tried to remove the electrodes. The electrode holder with the bee was rotated 90° and the bee was carefully adjusted onto the floating ball. The electrode holder consisted of a small balance which kept the animal on the treadmill with its own weight. This balance allowed the animal also to change the distance to the surface of the treadmill during walking. The direct light from the LCD projector was shaded in order to prevent direct illumination of the dorsal regions of the compound eyes and the ocelli. Two UV diodes were positioned within the shade just above the head of the animal simulating short wavelength light coming from above. The micromanipulator and binoculars necessary for positioning the tetrode during the preparation procedure were removed from the setup afterwards. Another manipulator allowed precise positioning of the animal on the spherical treadmill.

Drug application and Chemicals. Cocaine and octopamine were solved in DMF (Dimethylformamide). The solvent enabled a penetration of the insects cuticle and therefore external drug application to the hemolymph (Barron et al., 2009). Per treatment, 1µl Drug solution was administered onto the thorax. For Cocaine, doses of 3, 6 and 12µg per µlDMF were choosen, as they have been proven to show effects in honeybees (Barron et al., 2009). The same for octopamine with 3 and 6 µg per µl DMF. Ritaline was solved in sugar water, according to the doses used in *Drosophila* (VanSwinderen&Brembs, 2009, 0.5mg/ml). Controls were treated with 1µl pure DMF.

Analysis and Statistics. Spike2 (Cambridge Electronic Design) Software was used for data sampling and spike sorting. Before sorting, the data were high pass filtered. Matlab (2010b, MathWorks) was used for statistical analysis. Since the data presented here are not normal distributed (Kolmogorov-Smirnov test, lilliefors test), non parametric tests were used, like the Wilcoxon ranksum test to compare two variables or the Kruskal-Wallis test to compare more than two groups. If the Wilcoxon ranksum test was used to compare more than two groups, a Bonferroni correction was calculated. Therefore, the alpha level 0.05 was divided by the number of groups tested. For example 3 groups resulted in an alpha level of 0.016. The individual statistical relationships in the Kruskal-Wallis test groups have been analyzed with a multi comparison test. If the Confidence Interval (CI) of the multiple comparison test (based on the Kruskal Wallis test stats result matrix) does not contain zero, the groups are statistically different at the 0.05 significance level. If the CI does contain zero, there is no statistically significant difference for the tested groups at the 0.05 significance level (alpha). The mean plus standard deviations are shown for the spike rate data. Due to the high sample size in these cases, the mean plus standard deviation was a valid method to approximate that 68% of the data are in the interval of mean plus one time standard deviation, 95% of the data are found in the interval of mean plus two times standard deviations and 99% of the data are found in the interval of mean plus three times standard deviation. The two-sided Wilcoxon ranksum test was based on the null hypothesis that data in the compared vectors x and y (walking activity before drug application and walking activity after drug application, binned data) are independent samples from identical continous distribution with equal medians. H=0 indicates a failure to reject the null hypothesis, which means that the data in x and y are not statistically significant. The p-value was calculated for the 5% significance level.

Results Behavior Cocaine enhances the activity of stationary walking honeybees in a virtual environment. The activity was measured as rotation velocity. Additionally, for each bee the upregulation of activity due to the drug in comparison to the activity without drug application was calculated as ratio between the velocity difference (cocaine influenced running velocity minus baseline running velocity before drug impact) and the baseline activity (before drug impact) and is given as percentage. Three bees and one drone had statistically significant enhanced running velocities after the application of cocaine (bees 111212, 111214 (Fig.3, 4) and 111219: Wilcoxon ranksum 5% significance level: $p=1.8*10^{-8}$, 0.012, 0.0295, upregulation: 70-350%) (drone110930: Kruskal Wallis: $p=8.8*10^{-21}$, Confidence Interval (CI) [-55.1/-8.19], upregulation: +280%). In figure 2 the activity during night is shown as intensity map. The recording in 111214 started directly after the application of 6µg cocaine/1µl DMF at 10:22 p.m.. The walking activity increases approximtely 1 hour after drug administration during the first half of the night. Approximately 4 hours after the cocaine application a peak activity is reached. A decline in activity is seen after approximately 6h (fig.2). The same time scale is seen for the running velocity, calculated for the same animal and the same night (fig.3). In the second half of the night, periodic phases of inactivity are visible (fig.3). The same animal was recorded during the following day (fig.4). The recording started with drug application (3µg cocaine/µl DMF) at 11:17 a.m.. The animal shows enhanced walking speed approximately 1h after the cocaine application. Additionally, activity bouts are visible (periodic phases). In comparison to the night, the animal reaches higher peak rotation velocities during the day recording. Interestingly, the drones walking activity was enhanced with a dose of 12µg cocaine in 1µl

Dimethylformamide (DMF). In the forager honeybees an enhanced activity took only place in the case of 6 or 3μg Cocaine/μlDMF, not with higher doses.

One bee (111209, two-sided Wilcoxon ranksum test, 5% significance level: h=0; p=0.027) and one bumblebee (111130, two-sided Wilcoxon ranksum test, 5% significance level: h=0; p=0.13) have been tested with the highest dose of cocaine and did not show an statistically significant upregulation of walking activity after the application of 12μg cocaine/μl DMF in comparison to the drug-free walking activity. DMF alone did also not lead to upregulated walking activity (bee120114, two-sided Wilcoxon ranksum test, 5% significance level: h=0; p=1 (with a tendency for slower running, 50% downregulation of running speed with DMF in comparison to 0 treatment beforehand), bee 111209 (Wilcoxon ranksum test, with Bonferroni corrected alpha (0.05 significance level for the comparison of three groups=0.016 and control drone 2 (110930, multiple comparison test, based on Kruskal Wallis test result stats matrix, activity before and after DMF application: CI[-8.8/ 38.5], alpha level=0.05). As shown for the control bee, DMF application in control drone 2 led to a slight downregulation of the running activity (-32%) in comparison to the activity, measured before DMF treatment. Control drone 2 was not operated and therefore additionally used as control for the bees with implanted electrodes. After the application of 12μg cocaine/ 1μl DMF, the activity was statistically significant increased in control drone 2 (110930, multiple comparison test, based on Kruskal Wallis test result stats matrix, activity before and after cocaine application: CI[-55.1/ -8.2], alpha level=0.05). The upregulation of running speed due to cocaine was in the range of 280%, which is in line with results for animals with implanted electrodes.

Furthermore it was tested if beeboost (Queen pheromone) had an influence on walking activity. Interestingly, there is a statistically significant downregulation of walking activity (about -90%) with beeboost in near vicinity to the floating ball in control drone 2 (110930, multiple comparison test, based on Kruskal Wallis test result stats matrix, activity before and after cocaine application: CI[33/ 80], alpha level=0.05).

All three bees which have been tested with octopamine (6 and 2 µg/µl DMF) showed a statistically significant upregulation of running speed (111219, 111227, 111209: Wilcoxon ranksum: p= 0.0027, 0.0245, 7.75*10^{-4}, upregulation: 250-500%).

Ritalin alone did not lead to enhanced running speed (bee111207, two-sided Wilcoxon ranksum test, 5% significance level: h=0; p=0.6). After the application of slowly rising doses of cocaine (3µg cocaine/µl DMF at the first day (8:20p.m.), 6µg cocaine at the second day (10:30 a.m.) and 12µg cocaine at the second day (8:30 p.m.)) as well as one dose of 2µg octopamine (3rd day, 11:08 a.m.) Ritalin led to further enhancement of the running velocity (Wilcoxon ranksum: h=1, p= 0.0095, upregulation: 640%). The slowly rising dose of cocaine was statistically significant correlated with an upregulation of the running velocity (366%, two-sided Wilcoxon ranksum test, 5% significance level with Bonferroni correction for 3 groups, h=1, p=0.0072; (alpha=0.016). This cross medication effect of Ritalin was abolished if 12µg cocaine/µl DMF are directly applied prior to Ritalin (bee111209, two-sided Wilcoxon ranksum test, 5% significance level with Bonferroni correction for 3 groups (alpha=0.016): h=0; p=0.25) (111209, multiple comparison test, based on Kruskal Wallis test result stats matrix, activity directly before and after Ritalin application: CI[-5.2/ 18.9], alpha level=0.05). Interestingly, a treatment with octopamine in the same animal, between the cocaine and the Ritalin application led to a statistically significant increase in walking activity (bee111209, two-sided Wilcoxon ranksum test, 5% significance level with Bonferroni correction for 3 groups (alpha=0.016): h=1; p=7.75*10^{-4}). In bee 111209, the treatment started with 12 µg cocaine/1µl DMF at the first day (8:43 p.m.). The octopamine treatment took place at the second day (12:55 p.m.) and Ritalin feeding at the third day (3:28 p.m., 3µl).

The start of drug elicited activity is estimated by longterm recordings of walking activity in the virtual environment. The Cocaine impact starts approximately one hour after the application and lasts for about 4 hours (Fig. 2, 3, 4). Octopamine induced increase in walking activity seems to start later, approximately 6 hours after the application (Fig.5). Ritalin application leads to initially high walking activity but a long term reduction after approximately 1 hour (Fig.6). In Figure 7, a comparative long term recording of walking speed without drug application is shown. The walking activity is generally lower. After 2.5 hours in the same virtual environment the animal reduces its running activity drastically.

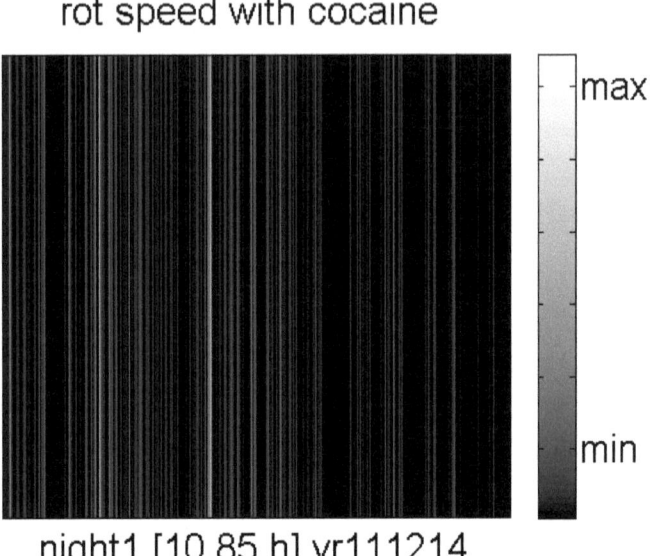

Fig.2: Night activity map for stationary walking bee 111214, directly after the application of 6µg Cocaine/µl DMF onto the thorax. The recording starts at 10:22 p.m.. The maximal walking speed is seen in the first half of the night. Enhanced activity starts approximately 1 hour after the drug application, reaches a peak after approximately 4 hours and declines after approximately 6h. High activity levels are indicated in light coloration (intensity scale on the right side).

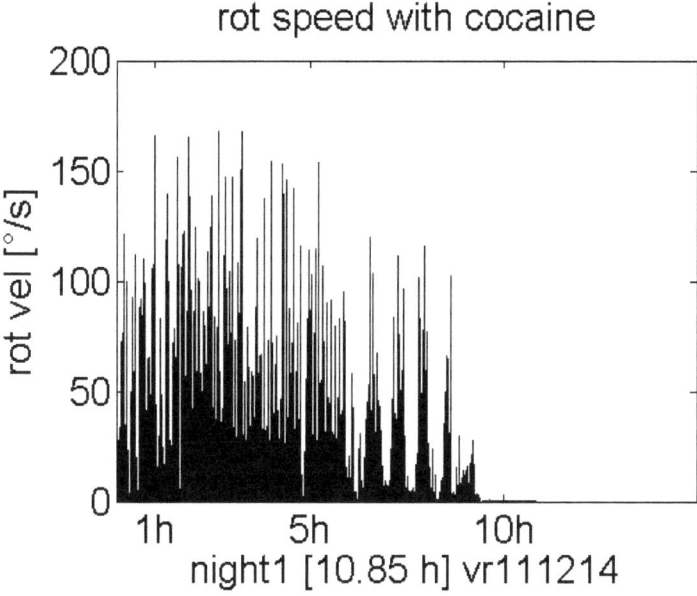

Fig.3: Rotation speed directly after cocaine application (6µg/µl DMF) during night starting at 10:22 p.m.. Velocity enhancement starts approximately 1h after drug application (start of the recording) and lasts for about 4h. In the second half of the night, periodic phases of inactivity are visible.

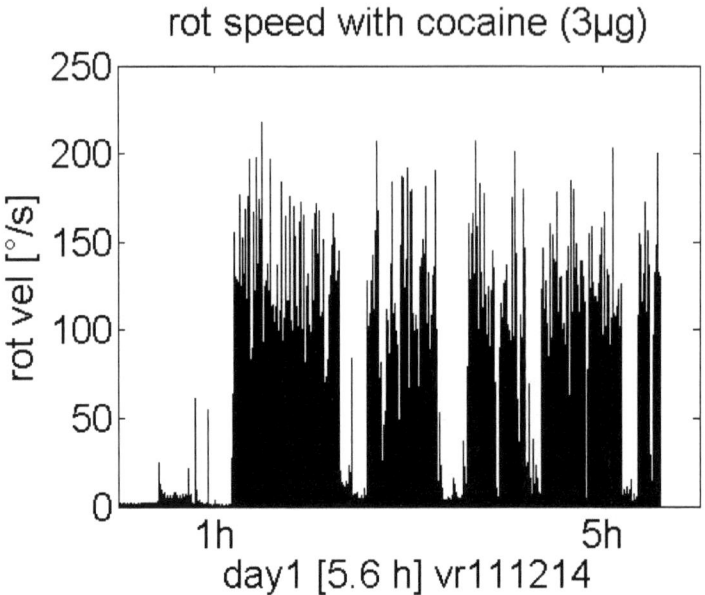

Fig.4: Rotation velocity at day 1, starting directly after the application of 3µg Cocaine in 1 µl DMF onto the thorax. The recording starts at 11:17 a.m.. The walking activity is enhanced after approximately one hour. Periodic phases of activity are visible.

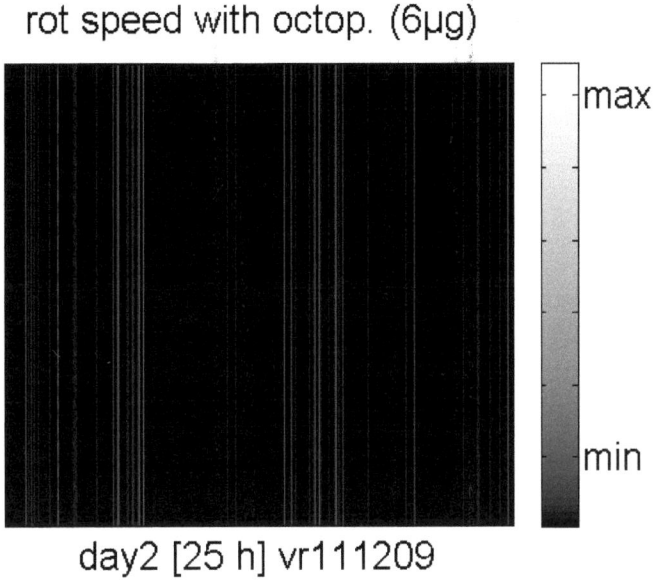

Fig.5: 25 hours activity recording immediately after the application of 6µg octopamine (12:55 p.m.). Activity enhancement starts approximately 6h after the application.

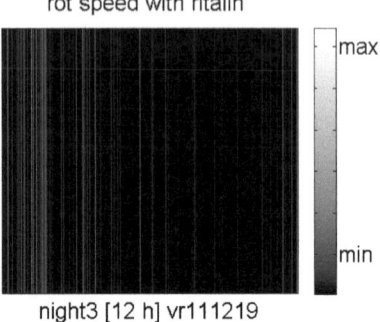

Fig.6: Recording of rotation velocity after Ritalin application (9:57p.m.). Enhanced activity starts after approximately 30 minutes and lasts for about 1.5 hours.

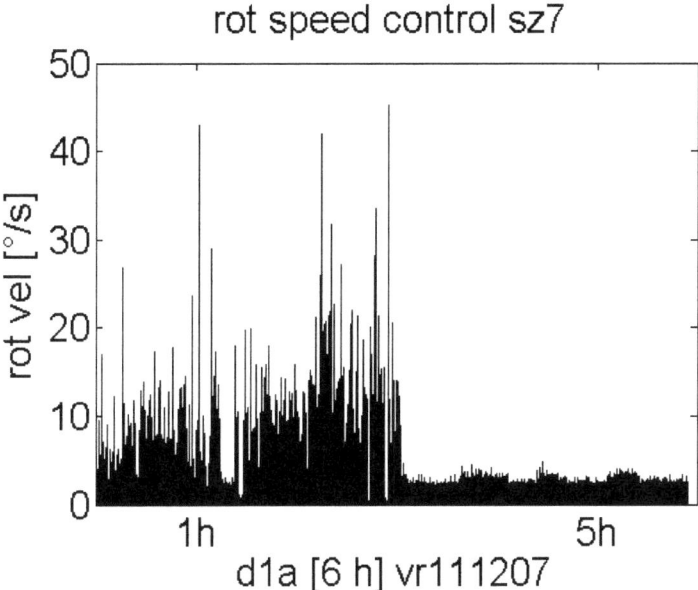

Fig.7: Running velocity in the virtual environment without drug application.

Results Electrophysiology. The spike rate in 111207 is increased in the range from 10 to 90 minutes after oral drug application. The baseline rate, measured 5h before Ritalin application (not shown) and 100 to 120 minutes after application of Methylphenidate (Ritalin), is very low with 1 spike in approximately 15 minutes (Fig.8). The spikes are easily identifiable by their special triplet form (Fig.9). During a period of 10 minutes after a second application of Ritalin, no further rate change can be observed.

In bee 111212, 5 units are recorded during stationary walking with visual feedback with and without cocaine. During the test situation without a projected pattern on the screen, the unit activity is initially high but without bigger fluctuations (sudden rate changes). The same is true for the walking activity (Fig.10) and for the first recording with a virtual szene (Fig.11). During the third test, a virtual navigation in the blender 3D sphere szene1, fluctuations in unit and walking activity appear Fig.12). Interestingly, unit 5 shows increased count per minute relative to the other units and in comparison to the other 2 control situations beforehand. Figure 13 illustrates a long term recording after the application of 3µg cocaine. Interestingly the units show very different reaction profiles. The units 1 and 3 (magnified selection of Fig.13 in Fig.14) show a steep increase in firing rate approximately 20 minutes after the application of 3 µg cocaine. During the plateau phase of high unit activity, these two units (which are recorded with different electrodes in the same region at the ventral border of the mushroom body) are highly synchronized. The synchrony of unit 1 and 3 is lost in the other low frequency phases, apart from the plateau. Unit 2 has no high plateau activity phase but some degree of synchrony to the early responding units 1 and 3. This first plateau phase lasts for approximately 20 minutes. After approximately 180 minutes unit 4 is activated in the same manner as mentioned for the early responding units 1 and 3. In addition to this later plateau phase unit 4 has a high peak fluctuation and seems to influence the rate changes of units 1 and 2. At the point when unit 1 and 2 activities are crossing each other, unit 3 activity is increased (Fig.13). Figure 15 shows a recording during virtual navigation in szenario7, immediately after the application of 6µg cocaine. Fast fluctuations in walking activity and unit spike count are visible. The units 1 and 2 show a high degree of synchrony (Fig.18). Higher rotation velocities are often preceded by increased spike counts for units 1 and 2 (Fig. 16, 17). In the following virtual szene1 (Fig.19) unit 5 has the highest spike count in comparison to the other units. This can not be due to the szenario, because in the following night recording with szene1, unit 5 activity is much lower again (Fig. 20). The activity relation might be antagonistically to unit 4 or with more complex correlations. Figure 16a demonstrates that the relation between unit 1 and unit 2 activity seems to influence the activity of the other units. 1.5h after the application of 6µg cocaine a sudden increase in unit activities 1, 2, 3 and 4 is visible (Fig. 21). Again, an increase in unit 1 and 2 activity is preceding phases of strengthened walking activity (Fig. 22). The units 1 and 2 show a single plateau of high spike frequency for a duration of approximately 90 minutes. The units 3 and 4 show a second plateau of increased activity for approximately 30 min during the end of the main plateau (Fig.23). Unit 5 does not develop a plateau activity but instead a short activity peak, approximately 2 minutes after the plateau onset of the other units (Fig. 23).

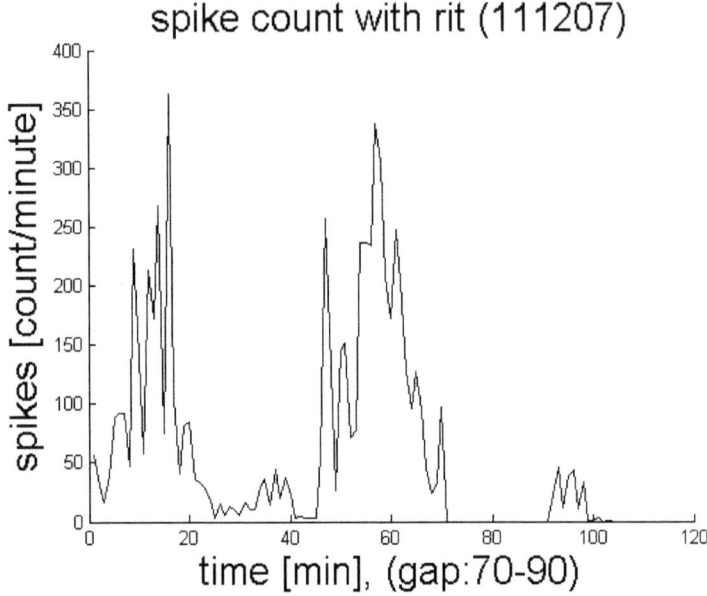

Fig.8: Spike rate from 10 to 120 minutes after Ritalin application. The recording has a gap between 70 and 90s, where no data are available. The rate 4h before Ritalin application and from 100s after the application onwards is 1 Spike per 15minutes. The recording site is the ventral mushroom body, PE1 region.

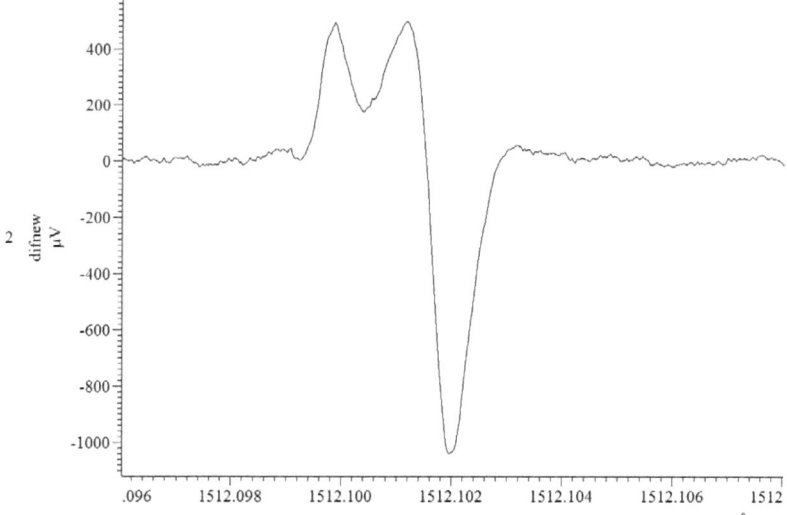

Fig.9: Representative Spike with triplet form, recorded from the ventral border of the alpha lobe in the PE1 region.

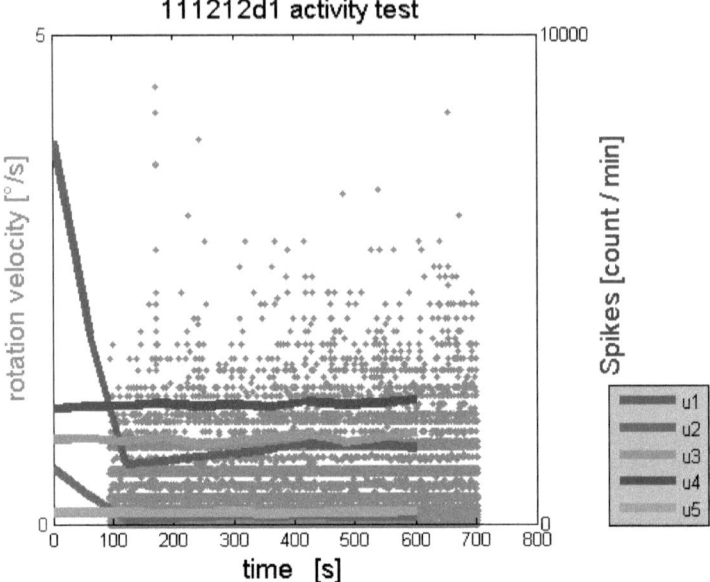

Fig.10: Bee 111212 in the activity test during stationary walking but without a virtual szene. All Five units show more or less stable activity patterns. Only in the start phase units 1 and 2 show higher spiking activity. This might be due to the handling before the start or the lowering of the cone and darker surrounding. Please note also the evenly distributed rotation activity (grey dots).

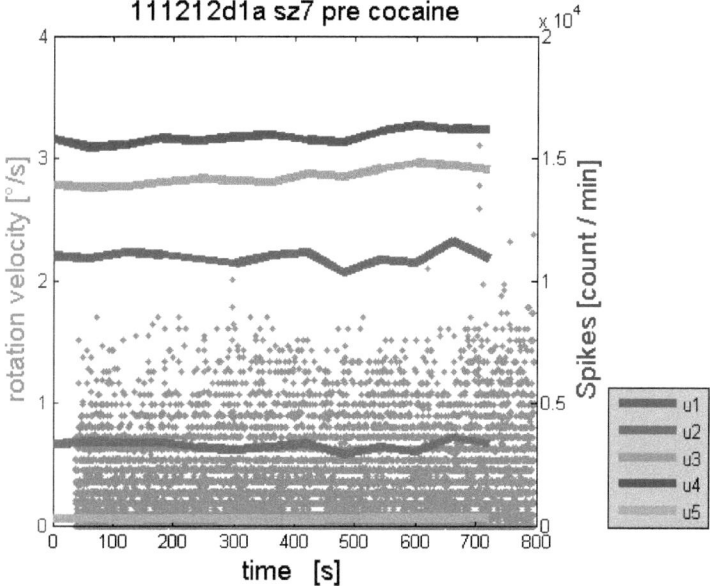

Fig.11: Walking activity and simultaneous spike patterns of 5 units in the virtual szenario7 before cocaine administration. The total amount of spikes is higher than in the walking test without a virtual szene.

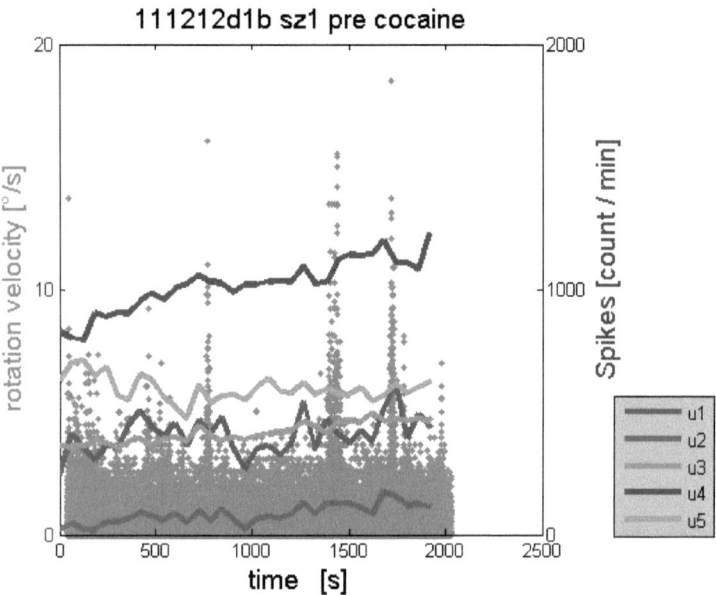

Fig.12: Walking activity and simultaneous recorded 5 units during virtual navigation in szenario1. The unit activity is lower than in the test and in szenario7 but there are activity fluctuations visible in the spike traces as well as in the walking activity. Please note the changed hierarchy of unit activity in comparison to the test situation with sz7. Especially the activity of unit5 is increased in relation to the other units.

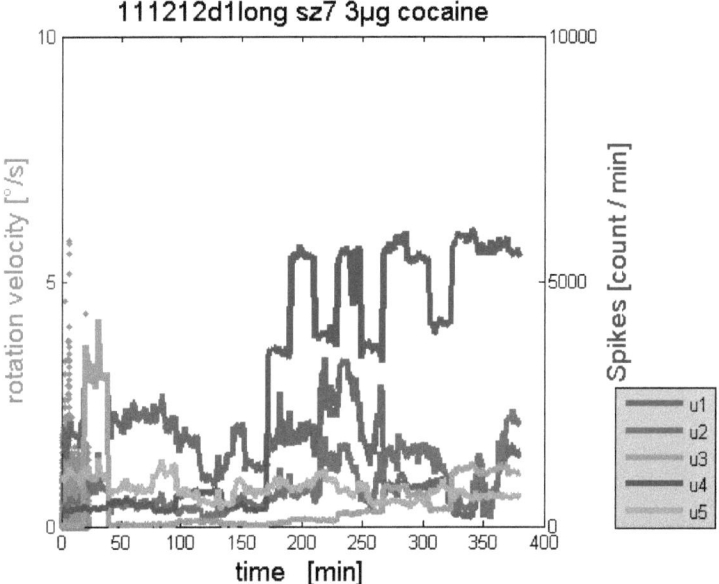

Fig.13: A long term recording of unit activity in bee 111212 after the administration of 3µg cocaine/µlDMF for more than 5 hours. A strong increase of unit 4 takes place after approximately 3 hours. It lasts until the end of the recording (approximately 3.5h). Strong fluctuations are visible. Units 1 and 2 are increased at the same point of time, whereas unit 3 is activated earlier and unit 5 does not show any pronounced increase in the firing rate. The rising phase of unit 4 is very steep.

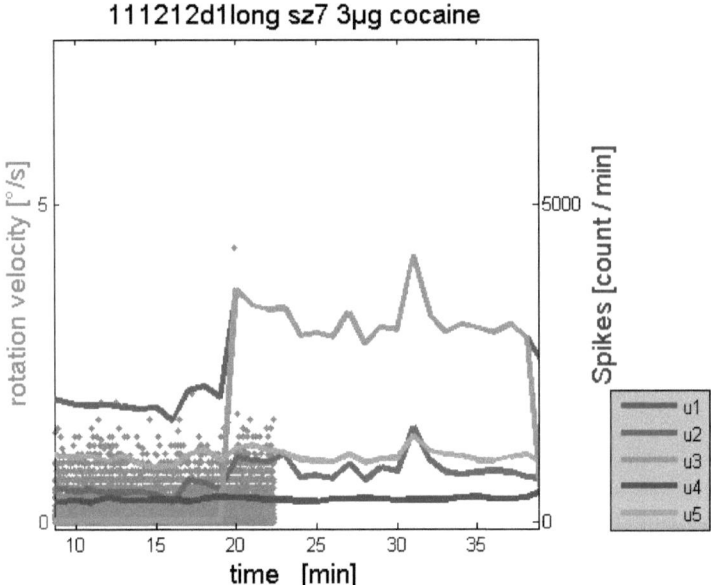

Fig.14: A zoom-in for the beginning of Fig.13. The steep onset of unit activity 1 and 3 (on different electrodes) is taking place approximately 2 minutes before the shutdown of the virtual szene (ending of grey trace).

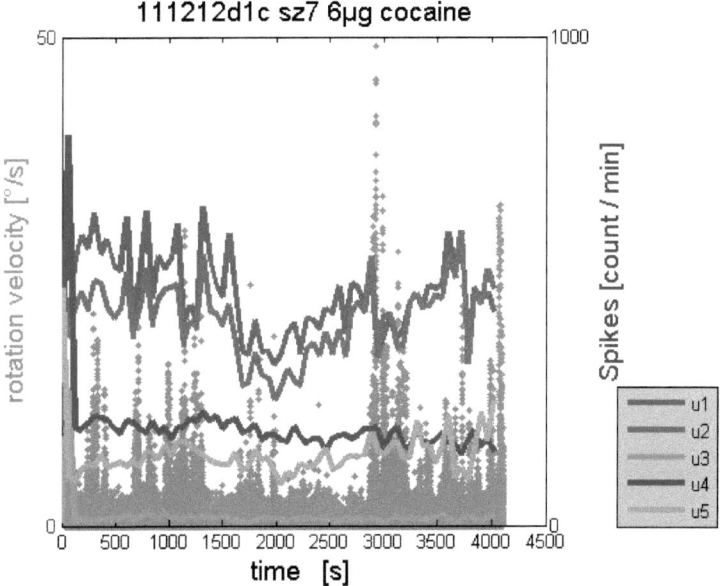

Fig.15: Neural and behavioral activity in the virtual szene7 after the application of 6µg cocaine per µl DMF. The units 1 and 2 are showing the highest activity as well as the biggest amount of fluctuations in the spike count.

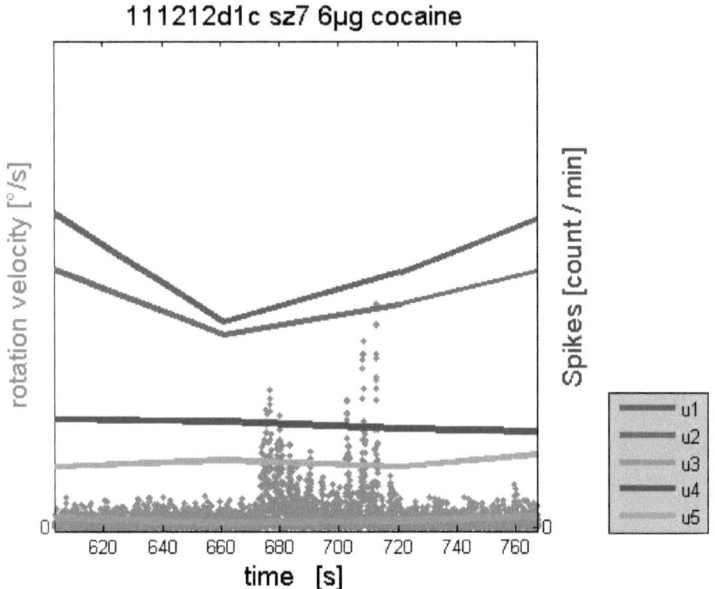

Fig.16: Zoom in to Fig.14. Spike count in the units 1 and 2 precedes motor activity.

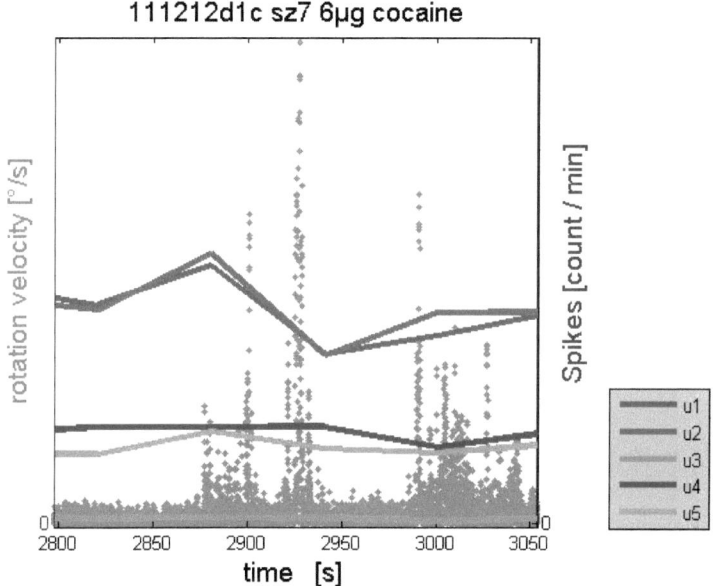

Fig.17: Second zoom-in for fig.14. A higher activity in the units 1 and 2 precedes motor activity.

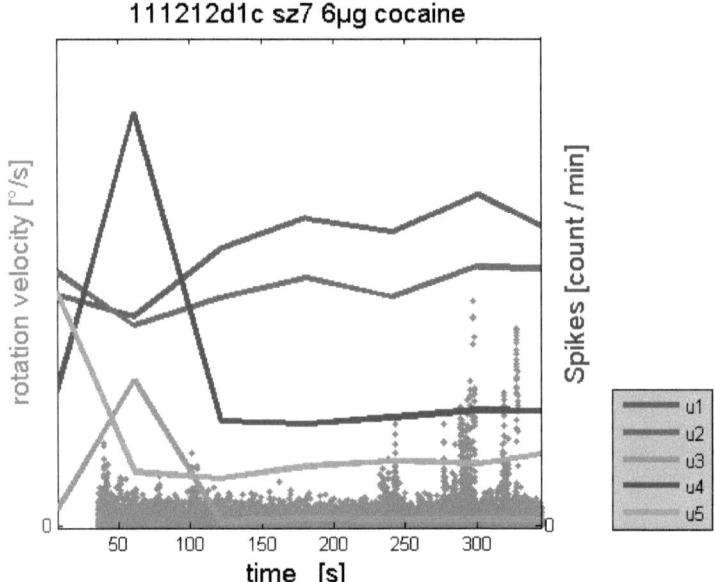
Fig.18: Example for dynamic correlations between unit activities.

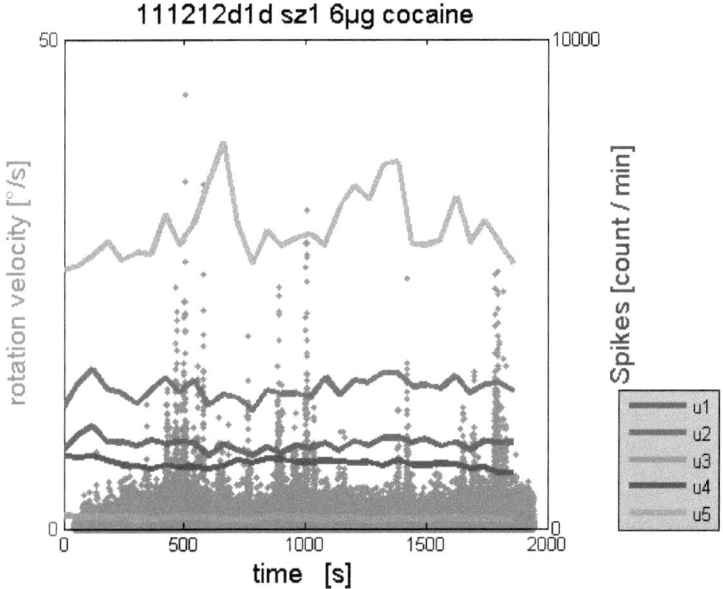

Fig.19: Spike count and behavioral activity during virtual navigation in szene1 with 6µg cocaine. Unit 5 activity is increased in comparison to all other units.

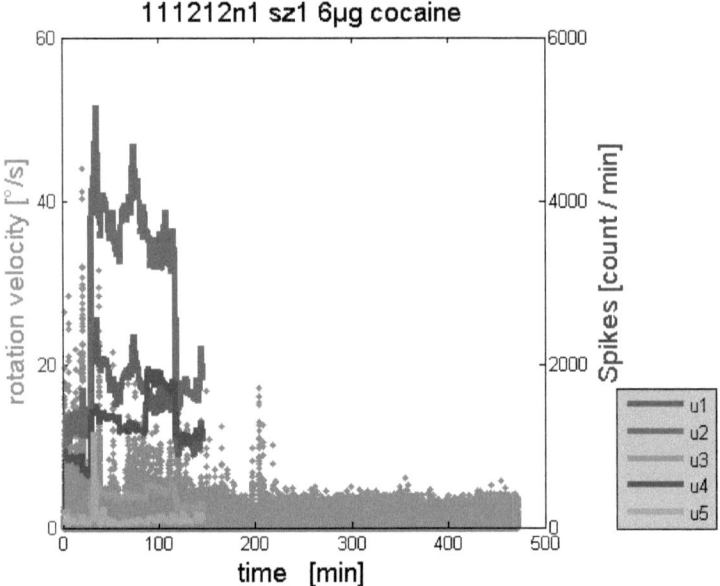

Fig.20: The strong and sudden increase in unit activity, 2h after the application of 6µg cocaine, lasts for about 1.5h.

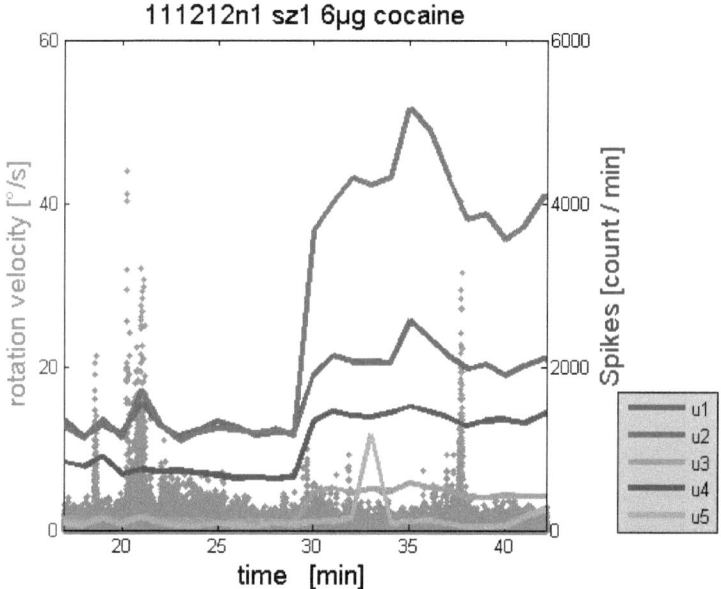

Fig.21: Zoom-in for the phase of sudden increase in unit activity, 1.5h after the application of 6μg cocaine. Units 1 to 4 show a synchronized onset in boosted activity. Unit 5 activity does not show a plateau but instead a clear peak in spike count, which is delayed for approximately 2 minutes in comparison to the increase in spike rate of the other units.

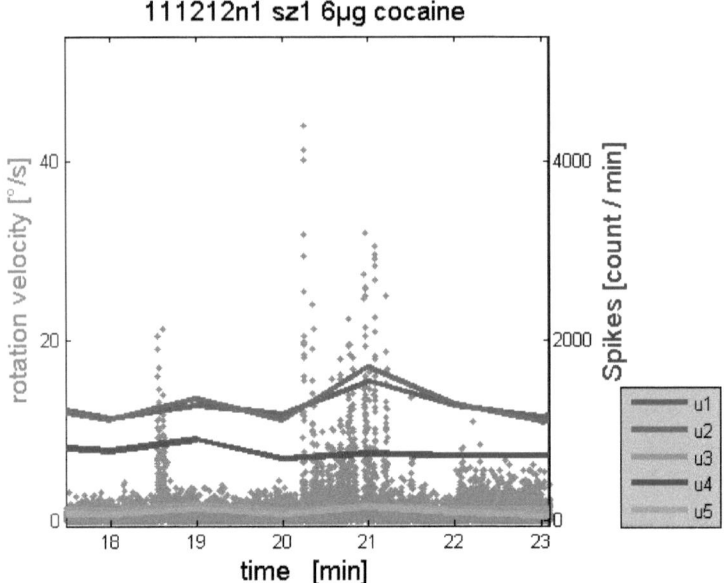

Fig.22: Phases of increased walking activity are preceded by rising spike counts of units 1 and 2.

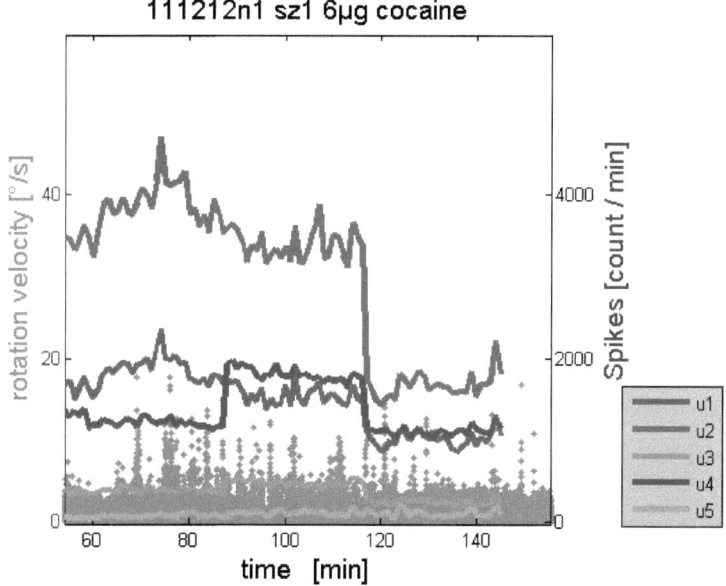

Fig.23: Synchronized ending of the plateau phase of higher unit activity after cocaine application during night. The units 3 and 4 show a second plateau which is absent in the units 1 and 2.

Discussion. It is impossible to simplify the presented results to one sentence of conclusion. Nevertheless, the main conclusion is, that the octopamine and dopamine reuptake inhibitors methylphenidate („Ritalin") and cocaine have complex and divergent influences on walking and neural activity and that the changes in neural activity are highly variable in neighboring neurons.

As predicted by the literature, cocaine and octopamine have an stimulating effect on locomotor behavior. Interestingly, for cocaine this effect seems to be non-linear and sex dependent in honeybees. High doses of cocaine (12 µg) lead to increased walking activity in drones but not in forager honeybees. Furthermore, the lack of locomotor stimulation by high doses of cocaine in forager bees can be abolished by gradually increasing doses. This effect leads to two assumptions. First, it gives evidence, that dopamine and octopamine transporters are somehow differntly organized in forager honeybees and drones. Second, it seems as if there is not only some form of addiction in honeybees, which results in impaired learning during withdrawal of cocaine (Barron et al., 2009) but also a form of dose adaption (development of tolerance) in forager bees. A recent study indicates that pretreatment with high doses of cocaine decreases the rewarding effects of the drug (Manzanedo et al., 2012). Furthermore, self-administration of cocaine leads to decreased dopamine transporter density in the nucleus accumbens of rats, whereas the dopamine transporter density is upregulated after self administration of methylphenidate (Calipari et al., 2012). It was shown that the dopamine reuptake inhibition potency of cocaine decreases during 5 days of self administration whereas the efficacy of methylphenidate remained unchanged (Calipari et al., 2012). Interestingly, methylphenidate alone did not alter the motor behavior in a forager bee but had severe influences on the neural activity in the mushroom body. This might be a hint, that Ritalin effects are more restricted to the brain than cocaine. Beforehand it must be clarified, if dose dependent or treatment (solving in sugar vs. DMF) dependent effects are masking similarities in cocaine and methylphenidate induced changes of behavior and spike rate. Neurons in the prefrontal cortex of rats show increased or (less often) decreased activity after acute methylphenidate application (Salek et al., 2012). The latencies and durations of increased activity are variable. The same kind of high variability has been observed for mushroom body extrinsic neurons after cocaine application (Fig.10-23). The strong activation of one unit due to Ritalin might be an effect of disinhibition via GABAergic feedback neurons (PCT) which partly express octopamine receptors (Sinakevitch et al., 2011). The activated unit has very characteristic triple spikes (Fig.9). Similar spikes have been observed in intracellular recordings in the PE1 neuron (Mauelshagen, 1993). The oscillatory activation might be due to inaccurate spike sorting and two independent units with diffenrent latencies. Nevertheless this possibility is very unlikely because the spike form is easily identifiable (Fig.9) and the homogeneity of the sorting result could therefore be controlled by eye. One possible eplanation might be that the „triple spikes" are three different units which show high synchrony but different patterns of activation after methylphenidate application. Another interesting point is the very low spike rate of this unit before the application of methylphenidate. 1 spike per 15 minutes is hard to detect and gives a hint for strong inhibitory influences on mushroom body extrinsic neurons which seems to be transiently abolished by Ritalin. It might be interesting to use Ritalin or other stimulating drugs to detect otherwise nearly inobservable, very sparse spiking units. In male honeybees, locomotor activity is increased with rising leves of dopamine during maturation (Akasaka et al., 2010) whereas rising levels of octopamine in maturating female worker bees are associated with the onset of foraging (Schulz&Robinson, 2001). These results indicate that the dopaminergic and octopaminergic system are differently regulated in drones and forager bees and may account for the dose dependent differences of cocaine to influence locomotor activity. The cocaine series (Fig. 10-23) gives another example for very complex network interactions and neural activity alterations after drug application in the bee brain. The five observed units are only a very small part of the network and therefore it is hard to draw conclusions from these observations. In the future it will be necessary, to develop new techniques to combine long-term and multi unit recordings in freely behaving insects, to gain further insights in drug related neural and behavioral changes. Simply due to the fact that the brain has the predominant task to change behavior in dependence of the environment, it is not surprising that the neural activity in the five recorded units is variable for different visual presentations without drug application (Fig. 10-12). The neural activity is increased with the onset of visual feedback (Fig.11). Another virtual environment (3D spheres, instead of stripes) leed to further increases in rate and changed relationships in the activity profile of the units (increase in u5). The application of cocaine leads to strong increases (plateau phases) in some units during day (Fig.13,14) and night (Fig.20,21). Interestingly, the activity profile is totally different during day and night. The day recording reveals higher variability and different latencies. During night, the plateau phase with high unit activity after cocaine application is synchronized apart from unit 5. Unit 1 and 2 show a possible relation to the walking activity (Fig.16,17). Some units develop oscillatory frequency changes (Fig. 13 u4) or dual plateau phases (Fig. 20, u3, u4), as it has been shown for the putative PE1 neuron after Ritalin application. Oscillatory or reverberating activity of mushroom body extrinsic neurons is a known phenomenon (Gronenberg, 1987) and might be enhanced after the application of stimulating drugs like cocaine or methylphenidate. It is surprising, that drugs like cocaine and Ritalin have similar effects on neural activity and behavior in insects as in humans or other vertebrates, since the biogenic amine system is differently organized. The differences are obviously more structural alterations than profound functional changes. Further, cocaine is much more stimulating on locomotor activity than Ritalin. In contrast, on the neural level, Ritalin seems to have severe impact on the spike rate. This might be in line with the use of Ritalin to treat hyperactivity disorder in children. The application of stimulating drugs seems to be a viable tool to detect sparse units and unravel the fascinating complexity of neural network interactions.

References:

Accardo, P., Blondis, T.A. (2001) What's all the fuss about Ritalin? J. Pediatr. **138:** 6–9.

Akasaka, S., Sasaki, K., Harano, K. and Nagao, T. (2010) Dopamine enhances locomotor activity for mating in male honeybees. J. Insect. Physiol. **56**: 1160–1166.

Agarwal, M., Giannoni Guzmán, M., Morales-Matos, C., Del Valle Díaz, R.A., Abramson, C.I., Giray, T. (2011) Dopamine and Octopamine Influence Avoidance Learning of Honey Bees in a Place Preference Assay. PLoS ONE **6(9):** e25371. doi:10.1371/journal.pone.0025371.

Barron, A.B., Maleszka, R., Vander Meer, R.K., Robinson, G.E. (2007) Octopamine modulates honey bee dance behavior. PNAS **104:** 1703-1707.

Barron, A.B., Maleszka, R., Helliwell, P.G., Robinson, G.E. (2009) Effects of cocaine on honey bee dance behaviour. J. Exp. Biol. **212:** 163-168.

Blenau, W., Baumann, A. (2001) Molecular and Pharmacological Properties of Insect
Biogenic Amine Receptors: Lessons From *Drosophila melanogaster* and *Apis mellifera.* Arch. Insect. Biochem. Physiol. **48:** 13–38.

Blenau, W., Erber, J., Baumann, A. (1998) Characterization of a dopamine D1receptor from *Apis mellifera*: Cloning, functional expression, pharmacology, and mRNA localization in the brain. J. Neurochem. **70:** 15–23.

Calipari, E.S., Ferris, M.J., Melchior, J.R., Bermejo, K., Salahpour, A., Roberts, D.C.S., Jones, S.R. (2012) Methylphenidate and cocaine self-administration
produce distinct dopamine terminal alterations. Addiction Biology doi:10.1111/j.1369-1600.2012.00456.x.

Chong, S.L., Claussen, C.M., Dafny, N. (2012) Nucleus accumbens neuronal activity in freely behaving rats is modulated following acute and chronic methylphenidate administration. Brain Research Bulletin **87:** 445–456.

Clark, M.C., Baro, D.J. (2007) Arthropod D2 receptors positively couple with cAMP through the Gi/o protein family. Comp. Biochem. Physiol. B **146:** 9–19.

Corey, J. L., Quick, M. W., Davidson, N., Lester, H. A. and Guastella, J. (1994). A cocaine-sensitive *Drosophila* serotonin transporter – cloning, expression, and electrophysiological characterization. Proc. Natl. Acad. Sci. USA **91:** 1188-1192.

Frölich, J., Banaschewski, T., Spanagel, R., Döpfner, M., Lehmkuhl, G. (2012) Die medikamentöse Behandlung der Aufmerksamkeitsdefizit-Hyperaktivitätsstörung im Kindes- und Jugendalter mit Amphetaminpräparaten. Zeitschrift für Kinder- und Jugendpsychiatrie und Psychotherapie, **40 (5):** 287–300.

Fussnecker B.L., Smith B.H., Mustard J.A. (2006) Octopamine and tyramine influence the behavioral profile of locomotor activity in the honey bee (*Apis mellifera*). J. Insect Physiol. **52:** 1083–1092.

Gallant, P., Malutan, T., McLean, H., Verellen, L., Caveney, S. and Donly, C. (2003) Functionally distinct dopamine and octopamine transporters in the CNS of the cabbage looper moth. Eur. J. Biochem. **270:** 664-674.

Goosey, M.W., Candy, D.J. (1980) The D-octopamine content of the heamolymph of the locust, *Schistocerca americana gregaria* and its elevation during flight. Insect Biochemistry **10:** 393–397.

Grohmann, L., Blenau, W., Erber, J., Ebert, P.R., Strunker, T., Baumann, A. (2003) Molecular and functional characterization of an octopamine receptor from honeybee (*Apis mellifera*) brain. J. Neurochem. **86**: 725–735.

Gronenberg, W. (1987) Anatomical and physiological properties of feedback neurons of the mushroom bodies in the bee brain. Exp. Biol. **46**: 115-125.

Hammer, M. (1993) An identified neuron meditates the unconditioned stimulus in associative olfactory learning in honeybees. Nature **366**: 59-63.

Humphries, M.A., Mustard, J.A., Hunter, S.J., Mercer, A., Ward, V., Ebert, P.R. (2003)
Invertebrate D2 type dopamine receptor exhibits age-based plasticity of expression in the mushroom bodies of the honeybee brain. J. Neurobiol. **55**: 315–330.

Kim, Y.C., Lee, H.G., Han, K.A. (2007) D1 dopamine receptor dDA1 is required in the mushroom body neurons for aversive and appetitive learning in *Drosophila*. J. Neurosci. **27**: 7640–7647.

Krause, K.H., Dresel, S. H., Krause, J., Kung, H. F. & Tatsch, K. (2000) Increased striatal dopamine transporter in adult patients with attention deficit hyperactivity disorder: Effects of methylphenidate as measured by single photon emission computed tomography. Neuroscience Letters, **285**: 107–110.

Lange, A.B. (2009). Tyramine: from octopamine precursor to neuroactive chemical in insects. General and Comparative Endocrinology **162:** 18–26.

Livingstone, M.S., Tempel, B.L. (1983) Genetic dissection of monoamine neurotransmitter
synthesis in *Drosophila*. Nature **303:** 67–70.

Manzanedo, C., García-Pardo, M.P., Rodríguez-Ariaz, M., Minarro, J., Aguilar, M.A. (2012) Pre-treatment with high doses of coaciane decreases the reinforcing effects of cocaine in the conditioned place preference paradigm. Neuroscience Letters **516:** 29-33.

Mauelshagen, J. (1993) Neural Correlates of Olfactory Learning Paradigms in an Identified
Neuron in the Honeybee Brain. J. Neurophysiol. **69:** 609-625.

McQuillan, J.H., Nakagawa, S. and Mercer, A.R. (2012) Mushroom bodies of the honeybee brain show cell population-specific plasticity in expression of amine-receptor genes. Learn. Mem. **19:** 151-158.

Monastirioti, M., Linn C.E.Jr., White, K. (1996) Characterization of Drosophila tyramine
beta-hydroxylase gene and isolation of mutant flies lacking octopamine. J. Neurosci. **16:** 3900–3911.

Ritz, M.C., Lamb, R.J., Goldberg, S.R., Kuhar, M.J. (1987) Cocaine receptors on dopamine transporters are related to self-administration of cocaine. Science **237(4819):** 1219-23.

Roeder, T. (1999) Octopamine in invertebrates. Prog. Neurobiol. **59:** 533–561.

Römpp, Lexikon der Chemie, Thieme, 10. Auflage

Salek, R.L., Claussen,C.M., Pérez, A., Dafny, N. (2012) Acute and chronic methylphenidate alters prefrontal cortex neuronal activity recorded from freely behaving rats. Eur. J. Pharmacol. **679(1-3):** 60-7.

Sasaki, K., Akasaka, S., Mezawa, R., Shimada, K., Maekawa, K. (2012) Regulation of the brain dopaminergic system by juvenile hormone in honey bee males (*Apis mellifera* L.) Insect Molecular Biology doi: 10.1111/j.1365-2583.2012.01153.x

Scheiner, R., Plückhahn, S., Öney, B., Blenau, W., Erber, J. (2002) Behavioural pharmacology of octopamine, tyramine and dopamine in honey bees. Behav. Brain Res. **136:** 545–553.

Schulz, D.J., Robinson, G.E. (2001) Octopamine influences division of labor in honey bee colonies. Journal of Comparative Physiology A **187:** 53–61.

Schulz, D.J., Barron, A.B., Robinson, G.E. (2002) A role for octopamine in honey bee division of labor. Brain Behav. Evol. **60:** 350–359.

Sinakevitch, I., Mustard, J.A., Smith, B.H. (2011) Distribution of the Octopamine Receptor AmOA1 in the Honey Bee Brain. PLoS ONE **6(1)**: e14536. doi:10.1371/journal.pone.0014536

Stern, M. (1999) Octopamine in the locust brain: cellular distribution and functional significance in an arousal mechanism. Microsc Res Tech **45**: 135–141.

Van Swinderen, B., Brembs, B. (2010) Attention-Like Deficit and Hyperactivity in a *Drosophila*
Memory Mutant. J. Neurosci. **30(3)**: 1003–1014.

Verlinden, H., Vleugels, R., Marchal, E., Badisco, L., Pflüger, H.-J., Blenau, W., Vanden Broeck, J. (2010) The role of octopamine in locusts and other arthropods. J. Ins. Phys. **56**: 854–867.

Volkow, N.D., Wang, G.J., Fowler, J.S., Fischman, M., Foltin, R., Abumrad, N.N., Gatley, S.J., Logan, J., Wong, C., Gifford, A., Ding, Y.S., Hitzemann, R., Pappas, N. (1999) Methylphenidate and cocaine have a similar in vivo potency to block dopamine transporters in the human brain. Life Sci. **65**: PL7–PL12.

Walker, R.J., Kerkut, G.A. (1978) The first family (adrenaline, noradrenaline, dopamine,
octopamine, tyramine, phenylethanolamine and phenylethylamine).
Comparative Biochemistry and Physiology [C] **61**: 261–266.

7. General Discussion.

Honeybees have most likely the neural requirements to build a cognitive map. It has been shown, that similar to rodents, ripple like events, replay-like patterns during sleep and place related firing („place cells") are detectable in the brain of hymenopteran insects. This finding leads also to the suggestion, that replay is an evolutionary conserved consolidation mechanism. Most likely, the last common ancestor of deuterostomia and protostomia, 600 Mio years ago, had the same mechanisms of spatial memory consolidation and cognitive mapping. It remains to be elucidated how exactly cognitive maps and place cells are working in invertebrates.

Furthermore, honeybees are able to transfer visual operant paradigms from a real world situation to a virtual one. Associated firing patterns are changing their burst mode, rather than the average spike frequency.

Ritalin and cocaine do not have the same behavioral impact, as shown in activity measurements of stationary walking honeybees. Ritalin as well as cocaine have severe impact on the frequency of recorded mushroom body extrinsic neurons. Furthermore, cocaine has also modulatory influence on network activity over longer periods of time. Interstingly, cocaine effects in honeybees are dose as well as sex dependent.

The data presented here are far too sparse and preliminary, to draw network conclusions and wiring schemes. Nevertheless, there are some prerequsites for hippocampal-like network properties in the mushroom body. Glutamatergic excitation from CA3 to CA1 region in the hippocampus is proposed to induce sharp waves (Taxidis et al., 2012, Maier et al., 2012). A significant population of Mushroom body intrinsic Kenyon cells express glutamate as neurotransmitter (Schürmann et al., 2000). According to a recently developed ripple model, interneurons get directly depolarized by

CA3 outputs and start spiking with high frequency. Further, the frequency range of these interneurons is adjusted by strong inhibitory feedback connections, which gives rise to ripple oscillations (Taxidis et al., 2012). The inhibitory role of GABAergic feedback neurons in the mushroom body of insects is an ongoing issue of debate (fig. D1, Grünewald, 1999, Gronenberg, 1987). Recently, phase locking has been shown for these neurons in a learning paradigm with honeybees (Filla, 2011). Only a minority of Pyramidal cells, which receive slightly more CA3 input than the other cells, overcomes the broad inhibition (Taxidis et al., 2012). This model is in line with the ripple cycle associated unit activity, presented here. The ripple associated units represent a minority of the recorded neurons. As shown here, stimulating drugs seem to partially overcome the inhibitory system in the mushroom body and provoke extensive firing in otherwise nearly silent units.

The two new methods, described here, allow for the investigation of navigation related brain mechanisms in insects. Nevertheless, it might be necessary to perform multi unit recordings with miniaturized versions of rodent microdrives (Kloosterman et al., 2009) in the future. Since the insect brain is very small, carbon fibres might be a suitable approach (Piironen et al., 2011).

However, the brain electrophysiology's „uncertainty principle" remains: It can never be excluded that the neural patterns under investigation are changed by the necessary influences of the measurements. I am sure, non invasive methods in electrophysiology will be possible in the future. One candidate method might be quantum entanglement, which is already used for quantum teleportation along remarkable distances (Ursin et al., 2004). Due to technical limitations, these quantum processes are presently hard to apply to classical systems. Photosynthetic organisms are a living proof that classical and quantum systems are not mutually exclusive. Photosynthetic active bacteria are using quantum coherence for light-harvesting (Strümpfer, 2012) since approximately 3.5 billion years (Nisbet&Nisbet, 2008). Hopefully an interdisciplinary field of Quantum Biology and Quantum Neurotechnology will emerge. This step might also be important for understanding the human brain and associated diseases. Probably, ethical aspects will be an emerging side problem.

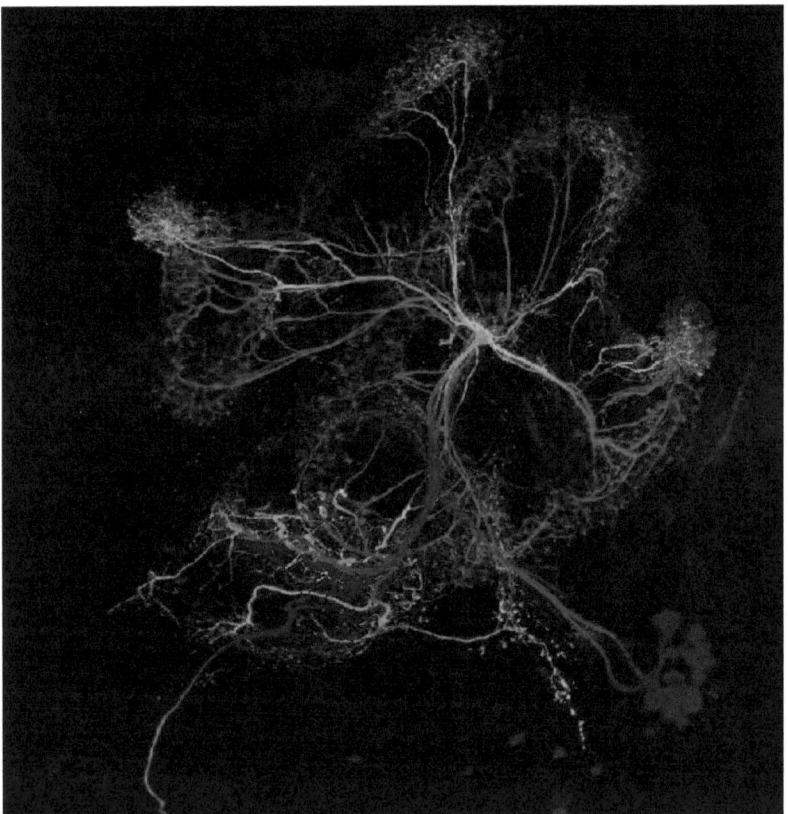

Fig.D1: PCT neurons in the honeybee brain (de Camp, 2009).

References.

De Camp, N. (2009) Struktur und Funktion der PCT Neurone im Bienengehirn. Diploma thesis, FU Berlin.

Filla, I. (2011) Associative plasticity and context modulation in GABAergic feedback neurons of the mushroom body output in the honeybee (*Apis mellifera*). Dissertation, FU Berlin.

Gronenberg, W., (1987) Anatomical and physiological properties of feedback neurons of the mushroom bodies in the bee brain. Exp. Biol. **46:** 115-125.

Grünewald, B. (1999) Physiological properties and response modulations of mushroom body feedback neurons during olfactory learning in the honeybee, *Apis mellifera*.J. Comp. Physiol. A **185:** 565-576.

Joesch, M., Schnell, B., Raghu, S. V., Reiff, D. F., Borst, A. (2010) ON and OFF pathways in *Drosophila* motion vision. Nature **468:** 300.

Kloosterman, F., Davidson, T.J., Gomperts, S.N., Layton, S.P., Hale, G., Nguyen, D.P., Wilson, M.A. (2009) Micro-drive Array for Chronic in *vivo* Recording: Drive Fabrication. JoVE.26. Http://jove.com/index/Details.stp? ID=1094, doi: 10.3791/1094

Maier, N., Morris, G., Schuchmann, S., Korotkova, T., Ponomarenko, A., Böhm, C., Wozny, C., Schmitz, D. (2012) Cannabinoids Disrupt Hippocampal Sharp Wave-Ripples via Inhibition of Glutamate Release. HIPPOCAMPUS **22:** 1350–1362.

Nisbet, E.G., Nisbet, R.E.R., (2008) Methane, oxygen, photosynthesis, rubisco and the regulation of the air through time. Phil. Trans. R. Soc. B **363:** 2745–2754.

Piironen, A., Weckström, M., Vähäsöyrinki, M. (2011) Ultrasmall and customizable multichannel electrodes for extracellular recordings. J. Neurophysiol. **105:** 1416–1421.

Schürmann, F.-W., Ottersen, O.P., Honegger, H.-W. (2000) Glutamate-Like Immunoreactivity Marks Compartments of the Mushroom Bodies in the

Brain of the Cricket. J. COMP. NEUROL. **418:** 227–239.

Strümpfer, J., Sener, M., Schulten, K. (2012) How Quantum Coherence Assists Photosynthetic Light-Harvesting. J. Phys. Chem. Lett. **3:** 536−542.

Taxidis, J., Coombes, S., Mason, R., Owen, M.R. (2012) Modeling Sharp Wave-Ripple Complexes Through a CA3-CA1 Network Model with Chemical Synapses. HIPPOCAMPUS **22:** 995–1017.

Ursin, R., Jennewein, T., Aspelmeyer, M., Kaltenbaek, R., Lindenthal, M., Walther, P., Zeilinger, A. (2004) Quantum teleportation across the Danube. Nature **430(7002):** 849.

Wardill, T.J., List, O., Li, X., Dongre, S., McCulloch, M., Ting, C.-Y., O'Kane, C.J., Tang, S., Lee, C.-H., Hardie, R.C., Juusola, M. (2012) Multiple Spectral Inputs Improve Motion Discrimination in the *Drosophila* Visual System. Science **336:** 925.

i want morebooks!

Buy your books fast and straightforward online - at one of world's fastest growing online book stores! Environmentally sound due to Print-on-Demand technologies.

Buy your books online at
www.get-morebooks.com

Kaufen Sie Ihre Bücher schnell und unkompliziert online – auf einer der am schnellsten wachsenden Buchhandelsplattformen weltweit! Dank Print-On-Demand umwelt- und ressourcenschonend produziert.

Bücher schneller online kaufen
www.morebooks.de

 VDM Verlagsservicegesellschaft mbH
Heinrich-Böcking-Str. 6-8   Telefon: +49 681 3720 174   info@vdm-vsg.de
D - 66121 Saarbrücken   Telefax: +49 681 3720 1749   www.vdm-vsg.de

Printed by Books on Demand GmbH, Norderstedt / Germany